CW00706539

Towards a Sustainable Economy

Towards a Sustainable Economy

The Application of Ecological Premises to Long-Term Planning in Norway

Stein Hansen

Pål Føyn Jespersen

and

Ingeborg Rasmussen

UHI Millennium Institute

11400545

First published in Great Britain 2000 by
MACMILLAN PRESS LTD
Houndmills, Basingstoke, Hampshire RG21 6XS and London
Companies and representatives throughout the world

A catalogue record for this book is available from the British Library.

ISBN 0–333–71520–9

First published in the United States of America 2000 by
ST. MARTIN'S PRESS, LLC,
Scholarly and Reference Division,
175 Fifth Avenue, New York, N.Y. 10010

ISBN 0–312–23229–2

Library of Congress Cataloging-in-Publication Data
Hansen, Stein
Towards a sustainable economy : the application of ecological premises to
long-term planning in Norway / Stein Hansen, Pål Føyn Jespersen, and Ingeborg
Rasmussen.
p.   cm.
Includes bibliographical references and index.
ISBN 0–312–23229–2 (cloth)
1. Sustainable development—Norway. 2. Environmental policy—Norway. I.
Jespersen, Pål Føyn. II. Rasmussen, Ingeborg. III. Title.

HC370.E5 H36   2000
333.7'09481—dc21
                                                                99–059245

© Stein Hansen, Pål Føyn Jespersen and Ingeborg Rasmussen 2000

All rights reserved. No reproduction, copy or transmission of this publication may be made
without written permission.

No paragraph of this publication may be reproduced, copied or transmitted save with written
permission or in accordance with the provisions of the Copyright, Designs and Patents Act
1988, or under the terms of any licence permitting limited copying issued by the Copyright
Licensing Agency, 90 Tottenham Court Road, London W1P 0LP.

Any person who does any unauthorised act in relation to this publication may be liable to
criminal prosecution and civil claims for damages.

The authors have asserted their rights to be identified as the authors of this work in accordance
with the Copyright, Designs and Patents Act 1988.

This book is printed on paper suitable for recycling and made from fully managed and sustained
forest sources.

10   9   8   7   6   5   4   3   2   1
09   08   07   06   05   04   03   02   01   00

Printed and bound in Great Britain by
Antony Rowe Ltd, Chippenham, Wiltshire

Contents

Preface

The 'Project for a Sustainable Economy' (*Prosjekt Bærekraftig Økonomi*) is a Norwegian research effort initiated shortly after the Rio Earth Summit in 1992 and completed in the summer of 1995. The project originated as a co-operative venture between Friends of the Earth–Norway (NNV) and the Project for an Alternative Future (PAF).[1] It later attracted the support of the Norwegian Ministries of Finance and the Environment, as well as the Research Council of Norway (NFR). The major purpose of the project was to conduct a critical review of existing long-term planning and modelling procedures, and to try to produce an alternative set of scenarios and projections. The feeling among the initiators was that current official projections relied too strongly on marginal adjustments in present production and consumption trends, thus maintaining the very development paths which had been seriously questioned at UNCED and in the follow-up sessions of the Commission on Sustainable Development (CSD).

The approach chosen was to employ the same economic models, policy tools and data being used in the official governmental projections. The aim was to retain the basic exogenous assumptions underlying the official projections (for example, population growth, technological development and the overall balance in public and current accounts), but to supplement these with more ecologically sensitive measures. Rather than accepting the premises and priorities of the official modelling procedure, the project turned to the technical division of Friends of the Earth–Norway for alternative perspectives and indicators. Taking on the role of a fictive 'Environmental Parliament', NNV prepared a report outlining the types of premises and indicators they would want to see applied if they had control over the public planning apparatus (Theisen, 1993). The demands of the 'Environmental Parliament' were then filtered through a series of 'consensus seminars' with modelling experts so as to determine which parameters could or could not be incorporated into the analysis.

Though some scepticism was voiced at the outset as to the feasibility of such an ambitious and unusual undertaking, there also emerged a growing political acceptance in Norway of the need for an integral treatment of key sustainability issues along with a mutual and open dialogue across the conventional barriers between environmental non-governmental organisations (NGOs) and the governmental actors responsible for environment-and-development policy.

Much of the original scepticism was based on a fear that any *serious* attempt to incorporate green demands into the planning and budget process would result in a radical decline in growth and the standard of living. Would the results of the project imply a long-term recession with massive unemployment and the end of the welfare state? Would Norway become an Third World Country? Would it mean having to part with luxury articles and exotic holidays for the average Norwegian, and what would the effect on the tax structure be? Would there be large-scale unemployment? Would ordinary citizens be given enough time to adapt to new demands and goals? Would Norwegian industry lose its competitive edge? And what about rural districts? Could they maintain the highly dispersed population structure which has become such a special feature of the Norwegian society? The questions – and the fears – were both numerous and strongly expressed.

Once given a forum for discussion, however, there gradually emerged an equal number of alternative questions. Did there not exist a major untapped potential for more sustainable energy and resource policies which were simply waiting for a stronger political mandate? Was there not a strong potential for changing the *pattern* of economic growth without immediate declines in the *level* of growth? Would not sustainable policies actually be able to create *both* new jobs and an improved environment? Would not sustainable policies allow more time for families to be together and for other quality leisure activities? Would not sustainable policies perhaps provide even greater security and welfare for the generations to come? Was there not a strong potential for new policies directed toward sustainable technologies and social solutions which were not currently 'realistic'?

These and numerous other issues were discussed as the design of the project took form, and was subsequently carried through. A major result of the consensus seminars was that only a limited number of 'green' dimensions could be directly incorporated into the current modelling apparatus. This part of the project ('green' GDP and macro-modelling) became the core focus of the research effort. So as not to gloss over the other 'demands', however, four themes were singled out for separate sub-project treatment: qualitative valuation of environmental goods; compensatory investments; environment and international trade; and environment and employment. The purpose of the present work is to relate the overall results from these five research efforts in a co-ordinated and (hopefully) instructive way. Part I of the book (Chapters 1 and 2) provides more detail on the premises and structure of the project and Part II (Chapters 3–10) treats the transition from premises to specific research demands – with all of the problems involved in achieving a more effective operationalisation. Part III (Chapters 11–15) presents the major findings from the four ancillary sub-projects, and Part IV (Chapters 16–18) the results for the macro-analyses and alternative projections. Part V (Chapters 19–22) then adds more specific alternative perspectives for agriculture, employment, transportation and fiscal policy, and Part VI (Chapters 23) rounds off the analysis with summary conclusions for the entire project.

Can it be said at the outset that the book presents an effective model for sustainable economic development? Probably not. It is, however, possible to hope that we have discovered something which might boldly be called a 'sustainable development strategy' – at least for the Norwegian economy. We have offered a perspective whereby, for a very modest 'insurance premium' in the form of reduced economic growth and changes in consumption and production patterns, emissions of climate-threatening gases can be significantly reduced. Only by attempting to combine the minimum demands for the preservation of life-supporting ecosystems with a secure transfer of the various components of common wealth to coming generations, can we claim to follow policies which *may* bring us closer to sustainable development. We believe that the Project for a Sustainable Economy, both methodologically and substantively, marks a significant step in this direction, and we look forward to a debate on the broader relevance of the analysis and results.

Any research effort of the size and complexity here involved generates an enormous amount of 'debt'. The Project for a Sustainable Economy is no exception. It can best be described as an enormous research jigsaw puzzle. In all, 10 private and public institutions and 17 researchers have, in one way or another, contributed to the project. The individual contributors to the sub-projects are acknowledged throughout the book, but there are numerous others, both in Norway and abroad, who have either provided written comments and input during the project's lifetime, or who have participated in the more than 100 presentations conducted by the project group thus far. The authors would like to extend their heartfelt thanks to all who have helped us in this way to achieve the critical dialogue we hoped for.

There are, however, certain individuals who deserve extra thanks. We would first of all like to thank the leadership and staff of the Project for an Alternative Future (and its successor, ProSus) for their ongoing 'domestic support' and feedback during the project period. We would also like to thank Fredrik Theisen (formerly of Friends of the Earth–Norway) and the research department of Statistics Norway, particularly Knut H. Alfsen, Bodil Larsen and Haakon Vennemo, for their vital contributions and patience during the work on the premises. Without their extra efforts, the project would probably never have come to fruition.

Last, but not least, we would like to thank the many sponsors of the project – the Norwegian Ministries of the Environment and Finance, the Research Council of Norway, and the various participating research institutes – who, in addition to contributing funds, have also actively participated in the project's reference group, providing useful comments and feedback throughout. They bear none of the responsibility for the conclusions here presented, but much of the honour for bringing the whole thing off.

STEIN HANSEN
PÅL FØYN JESPERSEN
INGEBORG RASMUSSEN

Part I

Background and Nature of the 'Project for a Sustainable Economy'

1
The World is Not What it Used to Be

1.1 The national perspective

Both the Norwegian economy and society have undergone great changes in the course of the past 40 years. The total wealth-creation power, measured in terms of the gross domestic product (GDP) in fixed prices has virtually quadrupled (see Figure 1.1).

In tune with this fourfold increase, there has been a virtually equivalent increase in energy consumption and use of material resources. During the period between 1950 and 1990, the total gross consump-

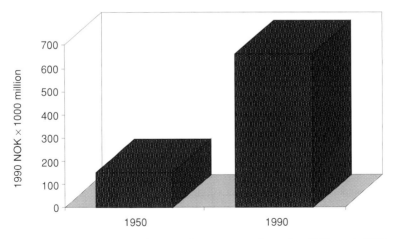

Figure 1.1 Comparison of gross domestic product, in 1950 and 1990, (fixed prices)
Source: Statistics Norway, *National Accounts*.

tion of energy more than tripled, from 190 petajoule (PJ) to 644 PJ, The mean annual growth in energy consumption during this period was 3.1 per cent, while the mean GDP growth for the same period was 3.8. Consumption increased most during the 1960s, when it showed an annual growth of 5.5 per cent. During this period energy consumption increased proportionally faster than GDP, while after 1975 the total growth in energy consumption was around 1.5 per cent. To put it another way, when compared with GDP the rate of increase in energy consumption declined somewhat until the mid-1980s. After this time, we see that energy consumption and GDP grow at virtually similar rates (Figure 1.2).

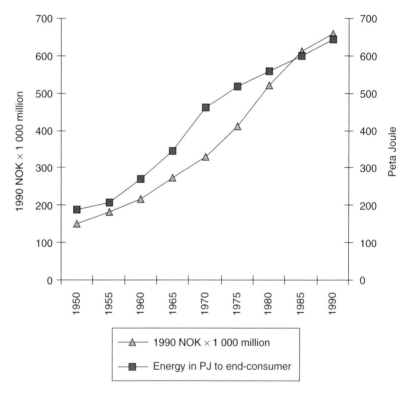

Figure 1.2.   GDP and energy consumption, 1950–90
Source: Bartlett (1993).

Norway has encroached upon more and more virgin territory to satisfy various human needs. In ever more areas, the country is beginning to realise that the environment sets limits for how, and how quickly, it can exploit these resources, at least if it is to be able to leave its descendants opportunities as great as those it inherited. In 1840, for example, there were a mere 15 000 km of public roads in Norway. In 1958, this figure had increased to 50 000 km, and by 1992 90 000 km, a nearly six fold increase between 1840 and 1992. During the same period, many industries and manufacturing sectors which were important and well established pillars of the Norwegian economy and society in 1950 have virtually vanished (for example, commercial whaling and timber floating), while completely new activities have assumed a dominating role (for example, oil and gas exploration, fish farming, computer technology and a wide range of public and private services). Women have also entered the workforce in a completely new and different way.

Structural change is an integral part of the processes of development and growth. Some of these changes can be traced back to changes in manufacturing techniques and market factors. When production and incomes increase, demand for certain products will increase more strongly than for others, quite simply because relative prices change during the process of growth, and because the products have differing income elasticities. Tastes also change over time, forcing subsequent change upon industry. The effects of these changes become apparent in many ways, including new production methods, further changes in consumption patterns, in the size and geographic distribution of import and export, in relative significance of industries, and in the social structure and distribution of incomes.

In virtually all industries (with the exception of health and care industries), modern production methods are completely different from those of 40 years ago. We see this very clearly in the reduction of labour needed to produce one production unit. In many cases, jobs have been replaced by more and better machinery and improved management techniques. The forestry industry is a clear example of this. The first revolution was the chain saw, then the forestry machines arrived. Lumber roads were built, and trucks removed the need for floating timber. The result of these trends can be seen in Figure 1.3.

At the same time as the new production methods are introduced, new materials are developed and brought into use. Synthetic fibres, plastic and carbon are examples of materials which have made new

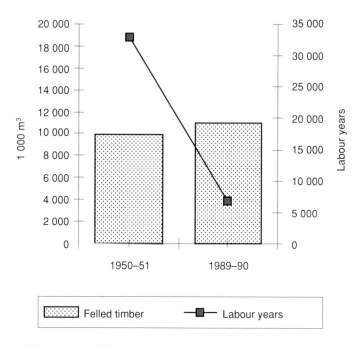

Figure 1.3   Forestry – felling and employment, 1950–1 and 1989–90
Source: Statistics Norway, *National Accounts* (1994b).

products possible, and which have resulted in the mass production, and new models, of previously familiar products.

Advances in production methods often lead to expansion of the product range; either by old products coming onto the market in new design or using new materials, or when completely new products create, and satisfy, new needs. The corporate giant, 3M, reports that 20 000 of its 60 000 retail products are newcomers, none older than four years (3M, 1994).

Tape recorders, cassette recorders, televisions and compact disc (CD) players have become household items since 1950. In 1951, there were 472 537 telephones in Norway, representing 15 telephones per 100 citizens. At the time of writing, 300 000 Norwegians are connected to the Internet, and the number of users will increase to 500 000 within two–three years. In a few years, in other words, Norway will have more PCs connected to the Internet than it had telephones in 1950!

Now think of all of the household machines we use in our homes, and even all of the sports and leisure products. How many squash

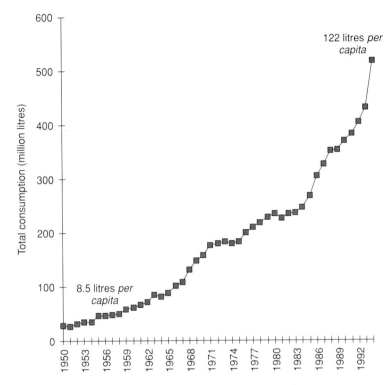

Figure 1.4   Consumption of soft drinks in Norway, 1950–94
Source: Soft Drink Industry Service Office ÅS (1995).

racquets were there per head of population in any country only 40 years ago?

New foods have arrived and new needs, or demands, change in tune with the new products. Making preserves and soft drinks from berries and fruit, or making one's own clothes, has become more and more rare. Frozen foods, microwave meals, pizza and pasta have entered the Norwegian food chain. Strawberries are cultivated in Norway all year round, and it imports exotic fruits from far-flung corners of the world. The consumption of soft drinks increased five times in Norway between 1950 and 1994 (see Figure 1.4).

New methods of production distort established related prices and influence the general public to change their consumption patterns. Mass-produced goods become relatively cheaper, while anything which requires human labour becomes more expensive. While former luxury

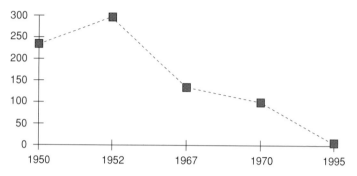

Figure 1.5   Telephone call charges for 1, minute Oslo to New York (1995 NOK)
Source: Norwegian Telecom.

goods become household items, domestic help becomes an unattainable luxury for most people. Figure 1.5 shows the price trend for a one-minute telephone call from Oslo to New York between 1950 and 1995.

Consumption patterns change as income levels rise. As people become more affluent, a smaller and smaller share of their income is spent on food and basic necessities and a greater share on travel, entertainment, leisure pursuits, and leisure articles. This trend has led to a gigantic increase in passenger transport (see Figure 1.6).

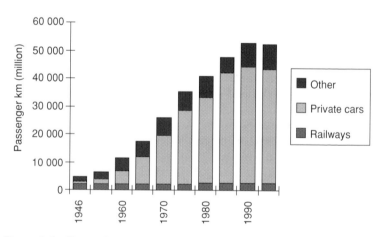

Figure 1.6   Domestic passenger transport, 1952–92, passenger km × 1 000 million
Source: Statistics Norway, *Historical Statistics* (1994a).

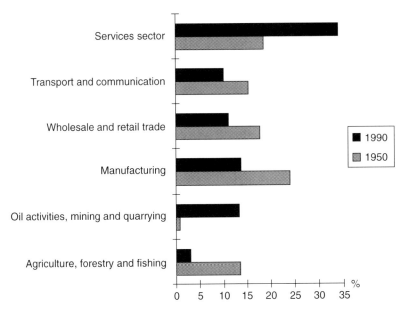

Figure 1.7   Selected industries' share of GDP, 1990 and 1950
Source: Statistics Norway, *National Accounts* (1994b).

When patterns of demand change, industry must respond by changing production patterns. Consumption of established goods and services changes its nature, and new goods, services and qualities must be allowed to join them in the market. This influences both manufacturers and raw material suppliers. For this reason, changes in consumption patterns and methods of production can make deep inroads into the structure of business and industry.

Developments lead to the network of interconnection between industries being gradually changed, with certain industries growing more rapidly than others. Production is transferred from the stagnating industries to those in expansion. Since 1950, the share of GDP contributed by the primary industries has receded, while service industries have shown a strong increase. We can also see that the share of GDP for both the heavy industry and the retail trade has decreased, while that for oil production has increased heavily. Figure 1.7 shows some of the changes.

The standard pattern of change for employment in the various industries is somewhat different. We can also see here that agriculture

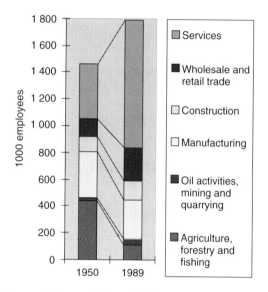

Figure 1.8   Jobs per industry, 1950 and 1989
Source: Statistics Norway, *National Accounts* (1994b).

recedes as income rises, and we see an equivalent heavy increase in service industries. Oil production has been of enormous significance for the GDP, although it is responsible for a relatively small percentage of jobs. All in all, 325 000 new jobs have been created between 1950 and 1989, but the total number of industrial jobs has decreased.

In fact, the real truth is even more dramatic, because the job figures for business and industry camouflage the displacement from production to service-related jobs within each individual industry. The relationship between workers and administrative staff has been strongly displaced, not just as a result of the increase in service industries, but also because traditional industrial companies are employing a steadily higher number of administrative and service staff per conventional blue-collar worker (Figure 1.8).

In addition to the displacement in number of jobs per industry, there has also been great change within the various industries. In the 1950 season, the whaling industry employed a total of 7500 workers, providing a total of 4000 labour/years. By 1989, no one was employed in that industry, whereas there were nearly 4000 labour/years in the fish farming industry. The greatest change however, was, the flow from the primary industries to the health and care professions. It is also worth noting that

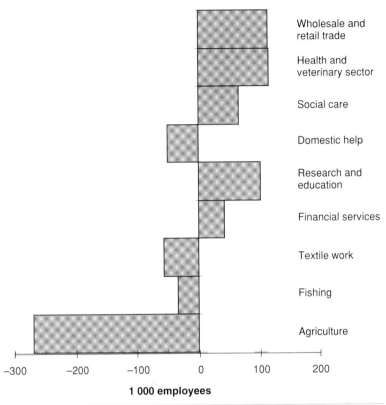

Figure 1.9 Workforce account – changes from 1950 to 1989, selected professions, (1 000 employees)
Source: Statistics Norway, *National Accounts* (1994b).

50 000 labour/years in the domestic cleaning profession have disappeared since 1950. Figure 1.9 shows the professions which have shown the greatest changes since 1950.

At the same time as the professions have changed character, the workforce itself has undergone radical changes in its structure. In 1950, women formed only 24 per cent of the workforce, while 40 years later they were more than 45 per cent. An interesting point is that the relative participation in the workforce by women fell from 1950 to 1960. The main reason for this was that the general welfare level rose, leading to increased marriage frequency and a lower average age of brides. In the 1950s, married women were less likely to work than single women. It is estimated that the workforce during this period was reduced by 38 000 as a result of women's change in civil status. From 1960 on, both married and single women joined the workforce in greater numbers, and women are now nearly half of the active working population.

Another significant change in the workforce is that fewer young people in the 15–24 year age range join the workforce as a result of the lengthening of compulsory education and increase in access to higher education.

1.2   The global perspective

Global changes have been even more dramatic (Box 1.1, 1.2 and 1.3). It is very tempting to compare it with exponential growth; in other words a trend which is hardly noticeable for thousands of years because there is a fixed, percentagewise negligible, change when the base rate is so close to zero. In time, however, these small growths, in percentage terms, become very noticeable as a result of the compound interest effect. A 3 per cent rise when an annual income is 1000 only amounts to a modest additional 30, while a 3 per cent rise when an income is 100 000 will be a more impressive 3000, or three times the entire annual income of someone who received the same percentage increase at the lower level. To put it another way: when an income has first risen to 100 000, the value of a 3 per cent growth in the course of one year equals an increase in value (with approximately the same resource use) which would require several decades when the principal was an income of 1000.

The first post-war years were characterised by the pressing requirement for rebuilding infrastructure and manufacturing capacity to be able to meet the needs of 2500 million people with relatively low real incomes (in 1950), and by little or no awareness on the part of the authorities or industries that nature had any limits of tolerance. The distance to these environmental boundaries was, however, so distant that unlimited withdrawal upon environmental accounts was taken for granted. Nor was the environment regarded as being a limited economic resource which required management, and for which payment was due if it was to be exploited.

By 1990, Norway's population had doubled, life expectancy had increased immeasurably, the real income had multiplied enormously, as had the use of renewable and non-renewable resources, with the associated emissions of waste products to air, water and soil. The prospect of this trend changing in the course of the next century seems to be most unlikely. More and more authoritative scientists have

Box 1.1   The changing global world

In 1950, who would have dared make the following claims?

- That in only a few decades there will be more Internet subscribers in Norway than telephone subscribers in 1950! No one would even have dreamed that something like the Internet would even have been invented, let alone be in general demand.
- That there would be a technological quantum leap in the early 1950s – the replacement of brittle 78 records by 45s and later LPs, which revolved far more slowly. No one, however, imagined the cassettes which later lead to the development of tiny portable tape players (the Walkman), and even later to CDs and the small portable CD players which are now a household item.
- That there would be words and ideas which are now part of the general vocabulary of all children in connection with high technology and information processing and exchange that were totally unknown in 1950 (at least, with their present meaning): video, cable TV, TV games, Nintendo, PC, laptop, Notebook, fax (telex will soon share the fate of the dinosaurs!), E-mail, diskette, hard disk, streamer, laser printer, ink jet, electronic time manager, desktop, photocopier, colour negative film, overhead projector, video and telephone conferencing, computerised cash registers, travel agents' booking systems, and so on.
- That there would be new, synthetic, strong, watertight, breathing, insulating, lightweight materials (carbon fibre, fibreglass, light alloys, Goretex), which are very often environmentally harmful. They have changed the art of industrial design, made entirely new products and markets possible, and thereby created new and different jobs!
- That there would be jet travel at today's discounted prices? For that matter, what about hydrofoils and hovercraft? Or even, driving your own car with ABS braking systems, seat belts and driver and passenger airbags as standard equipment. Nor was there anyone who had imagined studded tyres, not to mention catalytic converters.

Box 1.1 *continued*

Refrigerators and washing machines were familiar technology, although an unattainable luxury for most households, but who could have dreamt of microwave ovens, tumble dryers and dishwashers in every home? Many of the older generation can remember the first three-speed gear bikes. They were the cutting edge of high technology! In 1950, no one could have forecast the victorious arrival of 21-gear mountain bikes. Just think of all of the medicines and treatments which were generally accepted by medical science in 1950, and which are banned or discontinued today. Or fluoride in toothpaste and tablet form for children. These are developments which have affected many jobs.

We should also consider the activities and professions which disappeared, then reappeared, to the great surprise of forward-looking scientists: cycle couriers (although now with cellular phones and 21 gears), domestics/domestic services (often young people gathering work experience), and coopers at liquor importers. Traditional hand-made knitwear from Norway may, however, have originated in Sri Lanka or Bangladesh.

Who in 1950 would have dared predict that the whole of this extensive manuscript could be handed over to a publisher on 3 July, only to return less than six months later from the publisher as a beautifully bound, colour-printed, finished product!

If that's what we see today, we should be sensible enough not to try and predict what daily life will look like in 2030!

---

expressed their concern for the capability of the Earth to withstand this trend in such a way that options for future generations – even the possibility of the survival of the human race – will still be satisfactorily available.

Inequality between the rich and the poor has also increased. While the United Nations Development Programme (UNDP) calculated the ratio between the poorest 20 per cent of people on Earth and the richest 20 per cent to be 1:30 in 1960, the same ratio, and therefore the wealth gap, had doubled to 1:60 in 1990. Absolute poverty is also greater in many developing countries – especially African countries – than it was 30 years ago. Many Asian countries have managed to break out of the poverty trap.

Box 1.2   Disturbing global environmental trends

The report of the Brundtland Commission in 1987, its 'reconvention declaration' in the spring of 1992 and UNCED's Agenda 21 (1992), have all come to the conclusion that there are limits to the extent of material exploitation and deposition of waste and emission from human consumption that the Earth can withstand. There are especially nine areas which have come into focus over the past few years:

- The greenhouse effect and its potential for harmful climatic change
- The deterioration of the ozone layer
- The accumulation of environmental poisons
- Deadly acid rain falling on forests and killing marine life
- Pollution of the oceans
- Extinction of species
- Loss of topsoil
- Diminishing freshwater resources
- Forest clearance.

There is much uncertainty among researchers as to whether vitally important thresholds have already been crossed, or are in the process of being crossed. There is, however, broad agreement that present global trends are not sustainable, and that major ecological change can only be avoided if economic trends are drastically rethought.

This has given the North–South distribution problem yet another dimension; the ever-increasing income gap has been accompanied by ever-decreasing ability of the poorest people to resist and manage the change to the inhomogenous effects of rapid and irreversible man-made climatic and environmental change. This is the result of several factors. Among them, and perhaps most important, are economic policies which have ignored and displaced sustainable environmental management, increased population pressure and subsequently increased settlement in ever more marginal and exposed sites. The final result is to further escalate deterioration of the environment (Hansen, 1993a, 1993b).

Box 1.3   Poisonous chemicals

Between 1945 and 1985 the production of poisonous chemicals in the USA increased 15 times, to an incredible 102 million tonnes. We surround ourselves with around 70 000 chemicals in our daily lives and between 500 and 1000 new types of chemical are invented each year. The consequences and long-term impact of these chemicals are often unknown. Some of these environmental poisons can threaten human reproduction. A Danish survey of 15 000 men in 21 countries indicated that sperm quality had become 50 per cent poorer over the last 50 years. If these trends continue, then our ability to father children may be threatened in 20 year's time. Researchers think that artificial female hormones may be the cause of reduced sperm quality, an increase in testicle and breast cancer, and an explosive increase in the numbers of male infants born with deformed genitalia. This artificial oestrogen imitates or blocks production of natural oestrogen.

Nonylphenone is one of these oestrogen-like hormones. This is the surface-active material which produces an effective wash in washing powders, and it is also used in paint. The hormone is taken up and stored in human and animal fatty tissue, and has an extremely long degradation time. Each year, 75 000 tonnes are discharged into the environment in Europe alone. There are also 40 other products which are suspected of having an artificial oestrogen effect.

Source: Friends of the Earth–Norway, 1994.

As trends up to the present day have been marked by a seemingly exponential and accelerating process of change in business and consumption patterns, would it be reasonable for Norway to follow policies which attempt to satisfy the demands of various pressure groups, and thereby lock in place the structures of the 1990s? The answer to this question offers several partially competing dimensions: in the first place, it would seem that the process of freezing these structures in place will accelerate more and more, so that finally the country will quite simply not be able to afford conservation policies. In the second place, such freezing could be fundamentally opposed to the wishes and preferences of the general public, and there are probably limits on the amount that these can be controlled in a democracy. In the third place, freezing like this would lead to slower technological development within the energy industry and of environmentally sound products and manufacturing processes. In the fourth place, it would hardly seem to show solidarity when a structure which obviously requires an enormous welfare gap between the North and South is frozen. For the fifth, and this is not least the background for much of the present-day debate about sustainability, most people consider it environmentally unviable if the impoverished billions, who are today without cars and other luxury items which inhabitants of the rich countries consider to be completely natural, were given access to unrestricted use of these benefits. This last point focuses on redistribution of the fruits of the Earth in a more egalitarian manner.

The economical–political allocation and distribution questions which are connected to ever-increasing pressure on finite mineral, agrarian and recipient resources were raised to the status of a main feature in the Brundtland commission, which published its results in 1987. This attention culminated in the Rio conference of summer 1992 (UNCED), in the establishment of various global environmental conventions, in the establishment of the climate negotiations, in the Commission for Sustainable Development (CSD), and in the Global Environmental Fund (GEF).

Very few of the technical, social and economic structures we have today were predicted by future historians and forecasters in the early 1950s (Box 1.1). It may seem that the rate of change of externally given conditions, including those in the form of new technologies to which individual manufacturers and consumers have had to adapt, has increased. It seems to be clear that we must become accustomed to a future existence which is unpredictable, with stronger ties of dependency between local communities and nations than we have been used to.

At the same time it seems difficult to predict whether world-wide and local developments in Norway throughout the next decade will increase the need to determine a course of development which implements various measures to ensure for our descendants the same of level of welfare as we enjoy today. Experience leads us to believe that it takes time for changes in political opinion to be translated into measures which lead to changes in course. It is exactly because the strain on life-supporting ecosystems is increasing, and can suddenly prove to be irreversible, that it is vital that we protect ourselves through political decisions which ensure freedom of choice for the future through restraint in a number of fields today. It would be better for us all to enter a process of gradual change which would avoid us experiencing trends which lead to the inevitable destruction of all life on Earth than knee-jerk reactions leading to enforced measures after the irreversible destruction of the environment has been scientifically proven.

If the world community is effectively to combat the threat of global warming, the significance of products and services which cause appreciable emissions of greenhouse gases must be reduced. This means that national manufacturing industries and international transport, both sea and air, must be regulated. Use of such financial measures as price mechanisms would appear to be appropriate, since these would make metals and other materials and services which cause emissions much more expensive, signalling to consumers that they should begin to consider other materials or other ways of thought. It will scarcely be possible for governments to achieve regulation without taxation of fossil fuels and other harmful materials. The consequences for individual sectors are, however, so dramatic that no country has yet imposed CO_2 taxation on manufacturing industries. No country dares to be a forerunner in this process, because domestic opinion would regard it as synonymous with a desire to destroy the industrial base. While all nations wait for an international agreement and do nothing with the manufacturing industries or participation in international transport, the problem does not just persist; it actually increases.

It is against this background that PAF and Friends of the Earth–Norway have questioned the perspectives of the Norwegian government's Long-term Programme (LTP), which envisages only modest local emission reductions. None of these reduce CO_2 emissions compared with 1989 levels, but mean a virtual doubling of GDP and more than doubling of private consumption per inhabitant: in brief, trends in use of material resources which do not appear to accept that evermore people are competing for the Earth's limited agricultural resources and limited recipient networks for greenhouse gases from human energy consumption.

2
The Aims of the Project

Looking back on the long-term perspectives of social trends from the 1950s and 1960s, we can easily see that we were not very well equipped to forecast the future as far forward as 1990–2000. A great deal has happened that the prophesies did not envisage.

It is, however, still meaningful to attempt to illustrate how the growth potential and the composition of economic growth could be, given presumptions about technological trends and access to resources, and about the experience we have gained about the machinery of the Norwegian economy. Any such forecasts would, of course, be marked by being based on current conditions, and would only to a limited extent manage to include any of the genuine innovations we would expect to occur towards the year 2030. By looking to the future and applying present-day knowledge we should be able to outline the main trends and important features of developments which could be significant for present-day decisions.

Norway has long and unique traditions of long-term planning based on the philosophy outlined in Box 2.1.

Over decades of economic research, this has led to the development of operational planning models which describe how the Norwegian economy works, and how various steering measures can be used to influence developments. This analysis tool has, over time, become demystified and is familiar in political circles, in many ways forming a common basis for debate and proposals for the form of policies concerned with future trends.

As an example, consider the post-war growth of influential pressure groups who intend to protect and promote their own special interests. Measures they used, including macro-economic analyses which are based on this planning tradition, have resulted in much success in

Box 2.1   Development of national accounts and macro-economic planning

The first national budget was presented in the spring of 1946 under the title of *National Accounts* and *The National Budget*. The budget was to apply both for 1946 and for the 1946–50 period: 'the reconstruction period'. The second national budget, published at the end of March 1947 under the title of *About the National Budget 1947* planned only a year ahead, as did all of the subsequent budgets. Later, the expression 'long-term programme' was used for equivalent documents covering a longer period of time, first for the duration of the Marshall Plan (1949–52), then for the parliamentary sessions from 1954 to 1957.

At first, the budget concerned only the figures for the real economy – in other words the access to, and consumption of, products and services. The financial measures employed were not discussed in the first budget document and only partly in the second. The content was, however, expanded in time to apply to both the real and financial economies, and to all national economic or so-called *macro-economic policies* – that is, financial policy, monetary and credit policies, prices and income policies, currency policies and the general industrial policies.

In parallel with these developments, the central planning apparatus expanded from a small office with five employees to two departments at the Ministry of Finance, each with several offices, and a total of around 75 jobs in 1988. In addition to these were the jobs in all of the other ministries involved in drawing up the details of the national budget.

From the second national budget onwards, the planning was spread across the various ministries, and planning systems were, in time, put in place. The primary task for the Finance Ministry was to steer the planning process and co-ordinate planning proposals from other ministries. The annual planning process therefore became an integrated part of the normal activities of the state. Later on, the preparation of the Long-term Programme (LTP) was organised in the same decentralised manner.

Box 2.1   *continued*

In contrast to the state budget, the national budget and LTP applied to the total economy. This forms an important pillar for the proposals for the state budget, even if the LTP is not presented for the Norwegian Parliament at the same time as the state budget. In this way, the planning activities affect the state budget, which in turn affects the long-term planning.

During the drawing up process, and after the civil committees have considered them, the contents of the national budget and the LTP are discussed at several cabinet meetings. This enables the formation of the macro-economic policy to be systematised, so that both the basic figures and the professional analyses form a significant basis for decision. This applies both during the annual or four-yearly preparation of the policy document and when, at various points in time, major or minor policy changes are made.

The national budget and LTP have both verbal guidelines for economic policy and calculations of the most important items in the national accounts for one and four years. These figures are to a great extent arrived at today using a set of economic models, making it possible to create an economic policy which has a better coherence between objectives and means than was possible before the existence of such tools. Models like this had not been created when the first national budget was drawn up; the conceptions in the national accounts and the defined relationships between them, played a similar, if not so great, a part. They secured a logically coherent philosophy and gave a precise expression for the objectives and use of the tool.

The national budget is presented for Parliament together with the state budget in October each year. It is then discussed, first in the finance committee and then in the finance debate in November. Around six months later, the revised national budget is presented and discussed in a similar manner. The LTP is published separately before every general election, but is also discussed in both the finance committee and in a session of Parliament. There is no reason to believe that Parliament would be willing to refrain from this form of treatment of all macro-economic policy.

Box 2.1 *continued*

The national budget and the long-term programme play a part in political administration by, to a certain extent, compelling the government and civil service to follow the guidelines for economic policy laid down in them. The figures especially, make it easier to check later to what extent the processes in these planning documents have been adhered to. To this extent, the national budget and long-term programme seem to be a disciplinary measure. Deviation between the budget and the accounts does not have to be an indication of poor planning, nor even of poor execution of existing plans, although it could well be. If the preconditions for the plan change, then deviation is an expression of necessary modification.

During the immediate post-war years, there was great dispute about the national budget, and later the LTP, both outside and inside parliament. Somewhat later, the various governments from different political parties exploited these aids to economic planning; future governments of varying party colours will no doubt continue to do so. Nor have the planning documents been significantly changed whenever a new government has taken over the helm, although the contents and rhetoric have, of course, more or less changed.

Source: Bjerke (1989).

establishing permanent support and protection mechanisms, using subsidies, strategic market positions and protection against competitors. Primary industries and power-intensive raw material processing industry dominates in this development.

This was also useful for Norway as a nation in the immediate post-war period of structural adaptation. It has, however, become clear that the extent of the funding, and the way it takes place, inflicts both financial and environmental losses on the country, contributing to trends which cannot continue.

The Norwegian model tradition contributes to an explanation of why such private environmental bodies as Norges Naturvernforbund (NNV, Friends of the Earth–Norway) and such independent research projects as the Project for an Alternative Future (PAF) agree on an economically-based strategy for the protection of the environment when they wrestle with politicians over changing the influence of pressure groups. In other words, the same macro-economic arguments that pressure groups used to strengthen their position in the post-war period of reconstruction are now being used against them by the environmentalist organisations.

The ProSus research programme (Programme for Research and Documentation for a Sustainable Society) is a result of this process of political development. The initiative taken by PAF and the Friends of the Earth – Norway has brought environmental and social sustainability demands and goals to the macro-economic planning shops, where the economists create scenarios as a matter of routine using comprehensive macro-economic planning models. The ProSus project aims at simulation of long-term trends if the stringent environmental demands and goals which are regarded as minimum requirements by PAF and the Friends of the Earth–Norway for development to be directed towards a more sustainable course of development were to be imposed upon the Norwegian economy.

The aim of the project is, however, more ambitious than just the charting of alternative, intrinsically environmentally sound, courses of development for the future. The aim is to do this in such a way as to compare the calculated results with those arrived at as part of the perspective analyses of the LTP towards 2030. It is important to emphasise in this connection that the intention is not just to compare levels of polluting emissions in the project's and the government's alternatives, but also the impact of such drastic environmental policy restrictions on the balance of trade, the balancing of the state budget and employment figures. These are, of course, the figures by which the Finance Ministry and the Ministry of Trade and Energy set so much store. ProSus has chosen to use the government's own analysis tool and planning parameters for technological development, the international framework conditions, population trends, and so on, as the basis for the scenarios of this stringent environmental policy. When a common analytical platform is used, there should be a basis for direct communication between the government (and others who also consider these figures to be significant) and the environmental organisations about the consequences of the choices of direction which must be made in advance of the next four or so decades.

ProSus has engaged Statistics Norway (Research Section) to carry out some alternative simulations in the macro-economic models with other, far more stringent, environmental demands and goals than those used in the government's work on the long-term perspectives for the Norwegian economy. The project's desire to use these models is not based on a belief that they are especially suited to saying anything about the sustainability of a set of economic trends, but rather reflect a belief that the use of common model and analysis tools can contribute to easing communication between public bodies on the one hand and the green movement and the 'grass roots' on the other. Nor is the purpose of the model simulations which are presented here to produce detailed forecasts for the future, but rather to create a broader basis for discussion of present decisions in the shape of more, far less homogeneous, alternatives than the perspectives of the LTP. In areas where it proves that the authorities' model system is not suitable to illuminate possible trends, supplementary (if necessary, partial) sector or feature analyses will be applied.

The Project for a Sustainable Economy has had a very liberal attitude to the type of business structure which Norway should have in 2030. The project will not allow itself to make any particular business structure a goal in itself, nor will it make any concrete division of responsibility between the public and private sector such a goal. The project is primarily concerned with the strain on resources which results from selected solutions in the form of choice of technologies and consumption patterns. This focus arises from both national, regional and global concerns for resource management and equity between we who live today and the generations to come. If a constructive dialogue with decision-makers is to be made, it is necessary to answer more questions than just those about the environmental consequences of alternative development courses. It is also necessary to answer the question of whether we can afford (in the traditional, narrow budgetary sense) to follow the 'green' advice.

Part II

Towards an Operationalisation of Sustainability

3
Will 'Green GNP' Provide Better Resource Management?*

3.1   What is 'added value' growth and development?

The Project for a Sustainable Economy focuses on what added value is and what sort of future society we should aim at achieving. It is important to emphasise this at an early point in the book because inherent delusions about the nature of added value, growth and development seem to reign in a number of organisations, public bodies and companies, as well as throughout Norwegian society in general.

Value is created both in the manufacturing and service industries, in the public and private sectors. A modern economy is characterised by the total, mutual dependency between various industries and sectors, and this interdependence is often not evident until the total structure of both direct and indirect co-operation between various product and service producers and the consumers of these products and services has been included in the calculations.

It would be fatuous to claim that the industry manufacturing medicines and health care products creates value, while the health services (whether public or private) which ensure that medicine and care reach their targets, do not. It would be just as fatuous to claim that industry creates value when it manufactures steel and then processes it into scissors, if in the same breath you claim that the hairdresser who uses the scissors to cut a customer's hair – thereby creating demand for steel and scissors which enables industry to create value by manufacturing them – does not create value.

---

* This chapter, starting at Section 3.2, builds mainly upon Alfsen (1993). Section 3.9 also builds upon Asheim (1995b).

It is, therefore, futile to exclude value creation in a sector from the rest of society, as long as such interdependencies are the dominating characteristics of a modern economy. Value creation in business (primary, manufacturing industries and services) is dependent upon the efforts made in education, health care, construction, operation and maintenance of the infrastructure, in other words the services provided by the public sector (state, county or local authority). It is, in fact, tempting to claim that the establishment of social security and welfare schemes is not only intended to ensure people's welfare, but also to contribute to allowing them the financial risk which is necessary to attain socio–economic productivity gains through specialisation and division of labour.

Considering this allows us more easily to assess structural changes and altered conditions in the business world – for example, falling employment levels in agriculture and industry. Based on this, falling employment in the mining and heavy engineering industries can be a sign of health which indicates that the economy is in the process of moving in the direction of sustainability and increased welfare. Changes in employment, for example, are no ominous symptom of decrepitude, as spokespeople for industries in 'imminent danger of extinction' are wont to claim, while wielding the argument to defend special treatment and expensive artificial protection. The possibility of mechanising heavy work, and thereby liberating labour to more directly serve various human needs, should rather be regarded as a sign of health and strength.

In the light of this, it is incorrect to ask if we can afford care, an impression which can often be gained from political debate, where such questions are often reduced to the budgetary level, despite the fact that health and care costs have increased far more rapidly in real terms than the costs of other products and services. This primarily reflects the fact that productivity has been much weaker in the health and care sector than, for example, in industry. A continuation of the relative historic productivity and cost trends until 2030, which is the time perspective in the Project for a Sustainable Economy, would result in our using perhaps four times as much GNP on health and care measures than at present. The situation is, however, completely reversed when the focus is put on increased health and care costs at the same time as everything else has become so much cheaper. The actual productivity growth in the rest of the industrial base has formed the basis of the growth in real income, and these income levels have been matched in the labour-intensive health and care sector. As a result, prices rise without being accompanied by simultaneous improvements in productivity.

Is it sensible, however, to brand the steady increase we experience in products and services per working hour of other product and service production as a problem? After all, it doesn't mean anything more than that we liberate more and more working hours to more and better care. Box 3.1 illustrates the point:

Box 3.1   Affording care

Take as a basis for argument that it would be possible to care for 3 old people or produce 9 tonnes of aluminium with 1 labour/year. Then presume that after a few years it has become possible to produce 18 tonnes of aluminium with 1 labour/year, while it is still possible to care for only 3 old people. Can a problem have arisen when previously we could produce only 9 tonnes of aluminium and care for 3 old people using 2 labour/years, while we now are capable of producing twice as much aluminium, yet still care for only 3 old people using 2 labour/years? In other words, it is not a question of being able to afford care under such economic trends, but rather a question of the desire and will to place priority on care.

Source: Victor D. Norman 'Can we afford care?', *Dagens Næringsliv* (18 March 1995).

One of the core themes of our understanding of the functions of the economy is that our total resources are limited and that we must prioritise between them. In fact, it would seem that there is a widespread opinion that virtually the opposite is the case: we must expand the ever-more productive private sector to be able to expand the stagnated (in terms of productivity) health and education sectors and remove the associated queues and waiting lists which are a bone of contention for so many.

It is only when we blatantly and obviously waste resources that we can, with a large degree of security, claim that it is possible to have your cake and eat it, a so-called 'no-lose' situation. When there are high levels of involuntary unemployment and free, or heavily subsidised, access to a wide range of products and services, such wastage actually occurs. In this case, the challenge lies in transferring the economy to a non-cake and eat it situation – in other words, to a situation where increase in activity in one sector takes place at the expense of another. Then there is more to distribute to the community, especially making it possible to assist those needy people, who all agree are worthy, in a far more effective manner.

It must give cause for thought that most of us in modern society say that we are willing to give up many of the more or less useful consumer articles, if in return we gained shorter health queues, fewer corridor beds, less violence and crime, and a bit more time to enjoy life. For this reason, an important starting point for analysis of alternative future scenarios and a sustainable society is to admit that, in the long term, it is no solution to say that we must consume more of things for which we have, at best, limited interest, in order to receive more of what we most prefer. Policies of this kind would, in reality, remove us even further from the desired composition of the economic activities.

Bearing all of this in mind, we turn our attention to the national accounts and the indicators which are calculated from them.

3.2   GNP and other Expressions used in the National Accounts

Gross National Product (GNP) and a number of other collective markers or economic indicators are calculated on the basis of data from the national accounts. The UN has prepared a 'recipe book' for how to construct national accounts, under the title of SNA, an acronym for the System of National Accounts. There are many reasons for a country to wish to keep a set of accounts of its activities, but national accounts are primarily intended to provide a systematised framework and data to evaluate and analyse how the economy of a single country (for example, Norway) is functioning. In other words, to say whether it is doing well, whether there is a high level of business activity taking place, or whether the economy is in recession. Sometimes we wish to compare Norway with other countries, while at other times we can be interested in comparing Norwegian domestic trends over time. The national accounts are intended to cover all economic transactions in society, including, in principle, the illegal economy. National accounts also include certain products and services which are not retailed in the usual way in a market, where prices must be estimated; one example could be agricultural produce grown for own consumption on a farm or the value of living in one's own home.

A lot is also excluded from national accounts – for example, the value of housework in your own home, and certain aspects of what economists call 'external effects'. When two players (people or companies) trade, it will usually not affect anyone else than those directly involved: one of them exchanges money for a product or service and vice versa. Sometimes, however, such economic transactions can exhibit a positive or negative affect on their surroundings without their having participated in the deal. These side-effects are referred to as 'external effects'. One example of a (trivial) external effect is enrolling in an evening course in gardening. This will probably lead to a prettier garden, which will also benefit neighbours without their having had to participate in, or pay for, the course. While the evening course is a contributing factor to the national accounts, the benefit to the neighbour is not. Serious negative external effects are exhibited by polluting emissions and discharges to air and water, traffic accidents and noise pollution. To a certain extent, such conditions make us less economically efficient. Water pollution can result in periods of absence from work due to sickness; this will be registered in the national accounts as a reduction in economic activity. The pollution will, however, also lead to a welfare loss which will not necessarily affect our economic productivity, and this is not registered in the national accounts.

It is important to understand what is included and what is not included in the national accounts; in other words, the extent of estimated transactions in the SNA, not founded upon an economic or any other principle, but which are a function of the way in which the SNA is used. It is, therefore, difficult to direct only criticism in principle at, for example, GNP. The criticism must take into account the way the GNP parameter is actually used (or misused).

The national accounts also allow you to construct an aggregate of the economic indicator, GNP. A formal definition of GNP is that it is the total production of the economy in society as it appears in the national accounts, with the labour and material costs in the manufacturing industries deducted. In other words, GNP is a measure of value creation in the economy, and therefore of what is available for consumption, investment and export. One important use of our manufactured products is the maintenance of machines, buildings and other capital expenditure which is used for production. If we do not do this, our future ability to manufacture will be reduced. If the annual capital depreciation is deducted from GNP, we get what is known as the net national product (NNP). In many ways, this is a better measure of the values we actually have at our disposal in the long term, although it is somewhat weakened by a generally somewhat arbitrary valuation of capital depreciation.

Occasionally the expressions 'net national income' (NNI) and 'disposable national income' will occur. Net national income equals NNP *minus* what Norway receives in income from interest and returns from abroad. To put it another way, NNI is that part of the national income created in Norway; if we then also deduct Norwegian aid funds or donations to other countries, we arrive at the disposable national income – in other words, funds available for domestic use.

3.3   Use of the National Accounts and GNP

The most important application of GNP today can be briefly described under three headings: to follow economic trends over time, to study interaction between various parts of the economy and to compare international trends.

The national accounts allow us to follow trends in a number of economic parameters, including private consumption, investment, exports, imports, wages, taxes, and so on. It is also possible to calculate the value of various economic indicators – for example, net loans – to reveal how much is saved and invested in various sectors and the balance of trade, to name but two. These data make it possible to follow economic trends over time, and hopefully to give politicians and others enough information to make sensible decisions in financial policy-making. It is, however, worth mentioning that GNP alone cannot fill this role, but must also be supported by a number of other economic indicators.

National accounts can also be used to study the relationships between various sectors of the economy. The balance of trade, government deficits, inflation and unemployment are all significant expressions here. When economic models based on data from the national accounts are used, it is possible to study the economic effects of various economic policy proposals, for example, the question of how increased public sector activities would impact on unemployment, inflation and the balance of trade.

As most countries draw up their national accounts along the same lines (SNA), it is possible to compare economic trends for several countries, normally by comparing such figures as GNP or GNP per head of population. This type of information is often also used when fixing membership fees or contributions to such international organisations as the UN and the World Bank, or when the size of, and conditions for, international aid is fixed.

It can thus be seen that the national accounts in general, and GNP and other common indicators in particular, have several areas of use. In most cases, it is very easy to point out the deficits of the indicators. It would, however, be extremely difficult to construct indicators which could provide complete answers to all of the questions. On the contrary, GNP and a number of other indicators are designed to be compromises between the need for standardised, compact and plain information on the one hand, and sufficient relevance to the question in hand on the other. GNP is, in many connections, used as an objective and steering indicator for financial policy, despite it not being founded upon any economic or other principle. GNP is primarily used because it has shown itself to be useful in situations to be worth the trouble of calculating it.

3.4   GNP and welfare

The wealth of Norway is largely a result of the unusual concentration of such natural resources as hydropower, forests, fish, and not least, oil and gas. Norway is also, as it happens, a beautiful country with an environment so hugely unspoilt as to make it the envy of many. As Norway is in this situation, it is natural to ask if it is taking good enough care of the country's natural heritage. Could it be that the modern generation is frittering away petroleum riches at the expense of generations to come? Is Norway polluting the country 'too much' (whatever that means)? Is it really taking good enough care of flora and fauna? Many people have probably also found out that the economic progress that Norway has enjoyed during the past centuries has not led to an associated increase in quality of life, partly because of increased pollution of soil, air and water, and partly as a result of changes in social patterns which have led to isolation and broken marriages, and so on.

It should be obvious to everyone that the welfare of a society – in other words, how good a society is to live in – depends on a number of conditions which are not measured by the GNP or any other indicators in the national accounts. Significant factors which are excluded include social liberty, justice, access to education and health care, and – of course – the state of the environment. For this reason such economic indicators as GNP or GNP per inhabitant would not, on their own, be enough to say anything about welfare trends in a country over time, or about comparable welfare states in different countries.

Over the last few years, these and other conditions have led to demands that a GNP corrected for environmental factors should be drawn up – a 'green' GNP. This wish is based on the opinion that politicians and others who wish to participate in governing development in society use GNP as an important governing indicator, not least when their choice is between various policy proposals. If GNP is growing rapidly, policies are sensible and all looks well for Norway. If GNP growth is low or absent, things are not looking well for Norway and the proposed policies should be rejected. The 'green', or environmentally corrected, GNP is intended to encompass environmental harm and resource depletion which naturally follows increased economic activity. In this way, the 'green' GNP is better able to reflect welfare changes than the traditional GNP measure.

In addition to it being difficult to say anything about welfare trends in a society, there is also the fact that it is also, in principle, extremely difficult to measure welfare changes for one person over a period of time. Not only the surroundings change; people themselves change in complicated ways. In exactly the same way, it is difficult to compare the welfare of two individuals in a meaningful manner. It is probably a fairly trite observation to say that many inhabitants of developing countries seem 'happier' in their poverty-stricken lives than rich Norwegians do. Perhaps it is more important how we perceive our surroundings than how they are, purely objectively speaking? Or could it be that it is the relative differences in a society that are important? It becomes even more difficult when we want to say something about the welfare of a whole society. Should an increase in our welfare be just as important as it is to someone whose condition is far worse (who earns less, or is unemployed, is seriously ill, and so on)? We should be very careful not to simplify these situations to questions of the figures of 'green' or other indicators (Haavelmo, 1993).

3.5   Does GNP increase with traffic accidents, noise and pollution?

One of the arguments against GNP which is often raised is that when someone is taken to hospital as a result of, for example, a traffic accident, GNP increases. In the same way, GNP increases when polluting emissions to the air are cleaned. It would seem to be a paradox that accidents and pollution should lead to an increase in GNP when welfare benefits do not result from accidents and pollution! There are several points to note in this connection.

- First of all, it is reasonable to assume that when an accident has first happened, or the polluting emissions are taking place, there is a welfare benefit in being taken to hospital or having the emissions cleaned. That there is an increase in GNP associated with this does not have to be so wrong, even if GNP is taken to be a welfare indicator (which it is not).
- In the second place, it is far from certain that GNP increases in the case of admission to hospital. The accident victim may have had a job which created wealth, a job which the accident may prevent him or her from carrying out. This would lead to a drop in GNP. The question now becomes whether this loss is balanced by the wealth creation in the hospital when the doctors and nurses 'repair' the victim.

It is possible to go a step further and claim that if society were to suffer fewer accidents, we would not need to train and employ as many hospital workers. This would release workers and other resources which could be used in other forms of wealth creation. In other words, treating accident victims displaces other forms of wealth creation. All in all, it would seen reasonable to believe that a society which suffered few accidents would have a higher GNP than a comparable society with more, even if hospital work were to be included in GNP. A similar logic could be applied to pollution. A society suffering from little pollution, which would then not require application of resources on purification and other measures, should usually create more wealth than a comparable society which pollutes heavily, leaning on purification to avoid damage to health and other detrimental effects.

Taking expenditure on hospitals and pollution treatment, the so-called *defensive expenditures*, out of GNP will not necessarily give a better indicator of welfare change than GNP. If GNP was corrected for these expenditures the problem would not be restricted to that of double accounting, but also that of defining the extent of defensive expenditures compared with expenditure on consumption. Expenditure on purification is quite clearly defined as a defensive expenditure. But should expenditure on defence be included? Or on policing? In fact, even expenditure on food could be regarded as defensive expenditure against hunger. The factors which should be removed rapidly becomes a question of what the 'corrected' GNP measure should be used for, and especially what other information is available to illuminate the relevant question. In any case, in most cases it would be useful to have the traditional (uncorrected) GNP measure as a reference.

3.6   The significance of GNP in the political debate

It remains a fact, however, that GNP – or rather, changes in GNP – often appears in debates about, for example, environmental measures. The significance of GNP, or near-GNP indicators, in political debate should not be overestimated. Even if the intention is narrowed to only assessing economic trends, we see that other indicators can be just as prominent. At times like these, unemployment, interest levels and exchange rates are naturally regarded as three important indicators. However, having said this, it does not mean that GNP trends are without significance. It is not uncommon to meet the argument that this or that environmental measure is too expensive, referring to GNP, or even that GNP growth would be reduced by a certain percentage. Let us look a little closer at what is necessary if GNP is to measure changes in the state of the environment and exploitation of natural resources.

3.7   Valuating environmental benefits: a chronic headache

Most of the valuations in the national accounts are based on market prices – in other words, actual shop prices and other transactions. In a hypothetical ideal economy with no external effects, these prices would reflect the value of benefits, in the sense that the supply of products (in other words, manufactured items) would be exactly equivalent to the demand for them at these prices. To put it another way, people are willing to purchase all that can be manufactured at market price; no more, no less. When valuation is as uncomplicated as this, then there is a lot of sense in aggregating values from the national accounts for such as GNP.

Valuation of benefits which are not sold in any market – for example, environmental benefits – is more complicated. Even so, valuation is completely necessary for correction of GNP, or any other economic indicator, for changes in environmental conditions. Some of the problem with putting a price on environmental measures can be illustrated in Box 3.2.

All the above is meant to illustrate some of the problems faced when valuating environmental benefits. This does not necessarily mean that information about the willingness to pay or interest costs are uninteresting. It is, on the contrary, important that such decision-makers as politicians are provided with such information. This is why the the Project for a Sustainable Economy has had this reviewed; a summary of the results is presented in Chapter 7.

Box 3.2 Valuing environmental benefits

Let us consider a factory which uses the water from a river in its manufacturing processes, and which discharges polluting wastes into the river. Let us also consider that a town downstream from the factory uses the river as its source of drinking water. What is the value of clean water in this case? Seen from the supply aspect, the value of the water is the price of purifying it to an acceptable drinking standard. In this case we value the water by the expense of acquiring a clean drinking water supply. If the same situation is seen from the demand aspect we would say that the price that the town was willing to pay to acquire clean water would be the value of the water. These two valuation methods could give very different results, and it is not entirely clear which should be used in a correction of GNP.

The problem becomes even more complicated by the following conditions, which we consider in more detail in the discussion of compensatory investment in Chapter 6. If the factory actually purified the water, then its manufacturing costs would increase, probably leading to price rises for its products. This could, in turn, affect other prices, leading to changes in what is both manufactured and consumed. Finally, GNP itself would be affected. This is especially true if the measure is 'major', by which we mean that it affects large parts of the economy. General discharge reduction targets in a country could be one example of such a 'major' measure. In other words, it is not enough to just find the value of the water to correct GNP for the polluting discharge, you must also correct the traditional GNP measure as well. This requires a model of the economy, and the work of correcting GNP becomes a fairly comprehensive analysis of relationships in the economy. This is an analysis of another character than that which we normally associate with accountancy. The division between analysis and accounts is not always sharp, but put in vague terms we could say that the results of analysis will depend on more preconditions (often hypothetical and therefore controversial) than accountancy should really have to deal with. The same also applies, of course, if you decide to define the value of the water for the demand aspect.

It is not, however, correct for accountants to make controversial decisions about the value of environmental benefits and include such decisions, partly covertly, in apparently neutral information about trends in an environmentally corrected GNP. The information should appear through analyses where the preconditions and assumptions are clearly presented and discussed. If the conclusion reached is that correcting GNP for changes in environmental conditions is complicated, then there is perhaps hope that the situation is easier for natural resources which are traded in the market place. Should it at least not be possible to correct GNP for the oil and gas pumped up from under the North Sea each year? That is the next thing we shall look at a little more closely.

3.8   The significance of petroleum wealth for GNP

Analyses around sustainability have proved that a central element in the demand for sustainable development is to pass down our wealth intact to the next generation. (See Chapters 4–6 for a more thorough discussion.) In addition to overseas debts, tangible assets (machinery, structures and infrastructure) and human capital (knowledge and technical ability), wealth consists of natural capital. The question then is how large natural capital is. To take a concrete example, we will examine Norway's petroleum wealth, in other words, the value of Norwegian oil and gas.

The value of a national asset is usually calculated as the total of the discounted income from the asset. The value of Norwegian oil and gas can then be said to equal the current value of future income from oil and gas production. The value of the oil and gas (petroleum wealth) is thus dependent on future income, and is calculated on the basis of what we mean to be reasonable future price rates and production size. In practice, it has proved that people have very different opinions about what are reasonable price and exploitation trends. For example, Repetto and his colleagues at the World Resources Institute (WRI) have in some circumstances presumed that future prices will increase at a constant rate according to the so-called 'Hotelling Rule'. Others operate with a constant price. It is quite clear that the results – in other words, the estimates about the value of the deposits – vary widely.

Economists at Statistics Norway (Brekke *et al.*, 1989) have calculated the value of oil wealth based on official price forecasts from the authorities in various connections at various times in recent history. Figure 3.1 shows their conclusion.

The solid line shows the value of the oil and gas production, while the dotted line shows how the estimated value of the oil and gas deposits has changed in the period between 1973 and 1989. Most of the changes in the value from one year to the next are a result of the changes in price expectations. In several years they even exceed the changes in the GNP itself! The uncertainty surrounding future oil and gas prices is, in other words, so great that correcting the GNP for changes in the wealth would make the size of the traditional GNP measure virtually uninteresting. Nonetheless analyses such as this present useful information for politicians and others. On the other hand however, we find that the results of the analysis should not be simply included in the accounts and the indicators which build upon them. Correcting GNP for the oil production is not, therefore, as easy as one might imagine.

3.9 When is GNP or national income a suitable indicator of sustainability?

A reasonable interpretation of the expression 'sustainable development' is that we should not use more than our actual income. The question, however, is what exactly we mean by 'income'. The desire to correct GNP can be seen in light of this question. One definition of income

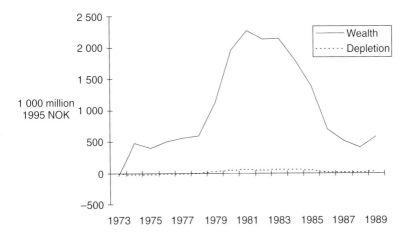

Figure 3.1 The variation in the value of the petroleum wealth against future price expectations.
Source: Alfsen (1993).

which has begun to be more common, and which is often ascribed to the English economist Sir John Hicks (1946), is as follows: *income is that part of the cash flow that we can use during, for example, one year without being in a worse financial situation than at the beginning of the period.* To put it more simply, we could say that in the traditional national accounts this would be the equivalent of the net national product – NNP or, in other words, GNP *less* the wear and tear on the real capital. The correction of GNP, or NNP, will then consist of subtracting the wear and tear on the natural and environmental capital – the attrition of capital. Some people would also claim that we should also subtract everything used to prevent harm to the environment (Daly and Cobb, 1989) but, as we have seen, this can lead to inconsistency and double accounting. It is, in any case, a difficult task to determine exactly what should be attributed to defensive expenditures.

We have also argued for the premise that it is extremely difficult to define, without controversy, what the value of environmental capital is. Calculating the wear on this capital is, of course, no less difficult and controversial. It has even proved difficult to find viable figures for the value of such a 'common' and internationally tradeable product as oil. In this case, the problems are closely connected with the fact that anything to be regarded as income is dependent upon what we believe to be the path of the future. When the expression 'income' is critically dependent on one's own expectations of future trends, then there is very little guidance as to how to act. To put it another way, if you are wrong, you are wrong – and even an environmentally corrected GNP would not tell you much more about the path to follow.

Apart from the problems in measuring living standards through this sort of consumption expression, there are several reasons why GNP, as measured at present, would not be able to fill such a function:

1. the capital attrition inherent in the exploitation of nature is not taken into account
2. the value increase in the remaining resources over time is also not taken into account.

We would like to now discuss what stringent and unrealistic preconditions should be applied if GNP or national income expression were to be a useful indicator of sustainability.[1] An important, although at the same time unrealistic, simplification in the 'green GNP' discussion is that one presupposes that there are market prices for all sorts of resources and capital holdings. The country is also assumed to have access to an international capital market where there is, at any time, a market interest rate which measures the return on investments in this market. Part of the country's capital holdings can be financial capital invested in this type of international capital market. GNP must take into account the return on such investment. We will also ignore the problems which the insecurity in resource price trends pose in the calculation of GNP (see Section 3.7).

As Asheim (1994b) has shown, there are very good reasons to distinguish between cases of constant market interest rates and cases where the market interest rates are expected to vary over time.

Scenario 1   constant market interest rates

The analysis is in this case is simple. The national income consists of:

consumption
+ the value of net investment in resource and capital holdings
+ capital gains

This means that both the wear and tear on capital involved in exploitation of natural resources and the increase in value of remaining resources over time must be taken into account. If sustainability is to be secured for successive generations, each generation must quite simply ensure that the market value of total wealth does not decrease.

Scenario 2   variable market interest rates

Theoretical models (for example, based on Solow, 1974) exist which show that each country, to a greater or lesser extent, has positive capital gains; but these positive gains are not included in the total income on a global level. The reason is that capital gain on a global level reflects reduced productivity, and cannot therefore, be regarded as a source of viable income. This reduced productivity means that the market rates are reduced over time (see the discussion in Chapter 5). A declining market rate does not, however, mean that we should ignore capital gains on a national level. They should be fully included, while at the same time taking into account the fact that the declining market rate means that the country must accumulate wealth if the trends are to be sustainable. The GNP must, therefore, be adjusted for interest effects. This adjustment will be proportional to the wealth of the country and will be negative if the market rate falls over time.

The formula for national income in this scenario would then be the following:

consumption
+   the value of net investment in resource and capital holdings
+   capital gains
+   the rate of change in the market rate times wealth

On a global level it can be shown that capital gains are positive if, and only if, the market interest rate falls over time (Asheim, 1994a). It can also be shown that if levels of consumption on a global level are constant, the total of the two final lines of the formula above will be zero (Asheim, 1994b). However, because the effect of interest rate adjustment is proportional to wealth, while capital gains normally are not, the two final lines are significant for the distribution of GNP between countries which are rich in resources and those which are not. Countries which are rich in resources and have large remaining resources, will have relatively large capital gains, so that the total of the two final lines will be positive. In the same way, countries with few natural resources, and a lack of remaining resources, will have relatively minor capital gains, so that the total of the two last lines will be negative. There is also reason to point out that – when there is a falling market interest rate – it is not possible for each generation to secure sustainability by ensuring that the market value of total wealth does not decrease. On the contrary, each country must increase the market value of its wealth over time to counteract the decreasing rate of return. It is exactly this aspect which is reflected in the adjustment for change in market interest rates discussed above.

3.10   Life goes on without 'green GNP': proposals for a strategy for sustainable resource management

What can be done to make decision-makers and others take sufficient care of environmental and natural resources in their activities? Our proposals for an answer can be divided into two parts:

• First, information should be provided about what has actually happened to environmental and natural resources in the past. One way in which this can best be done is to show the development in natural resource reserves in physical units. Priority should also be given to finding a set of environmental indicators which in a clear and unambiguous way can give information about the state of the environment in the country. This information, based on data in physical units of measurement, should be able to give a useful basis for decision for politicians when they face major problems within resource and environmental management. In brief, the advice is to include information in the accounts about physical natural resource and environmental conditions.

- In the second place, analyses should be produced, based on economic and other models, about how various policies proposed can affect economic growth, the form of the growth and the stores of natural resources and environmental conditions. The goal of Statistics Norway is to base these analyses on information about the physical effects of the deterioration of the environment on the working of the economy. One example could be the effect of air pollution on the level of absence through sickness in the workforce. If this is included as a parameter in the economic model, it will calculate what the value of a lost working day means to Norway. This approach avoids having to include more or less coincidental and controversial valuation problems, while at the same time excluding important aspects of environmental benefits. Cleaner air does not just give us less absence from work, but also other benefits on which it can often be difficult to put a price. Other analyses can raise the problem of putting a price on petroleum assets, and how it is dependent on the factors which Norway believes will affect future oil and gas prices. This will contribute to telling us something about whether we use more than the return on these assets at present; whether we are actually tarnishing the 'family silver'. This sort of analysis will always be controversial, as it builds on more or less well-based estimates about the valuation of environmental benefits and natural resources. However, the analyses will be able to illuminate the uncertainty and the debate around the central presumptions in a completely different manner than the figures in a set of accounts. In other words, we believe that neither the environment nor the social debate are well served by our forcing all of this information into a single figure, presented as a correction to another figure.

Of course, these proposals force politicians and other decision-makers to face a more difficult task than assessing a single indicator – 'green' GNP. On the other hand, it may contribute to making politicians more aware that the challenge of ensuring sustainable development is more comprehensive than assessing the value of one indicator compared to another. You have to face up to the reality, and all of the uncertainty, connected with future trends. A 'green' GNP will, in many situations, contribute more to hiding problems than solving them.

The Project for a Sustainable Economy has taken up the gauntlet and wants to show how physical environmental demands, drawn up against the background of goals of a sustainable development, can be connected to the economic development indicators within the framework of the analysis apparatus that the state uses in the development of the perspective analysis of the long-term programme (LTP). However, in the Chapters 4–8 we shall present a brief review of some of the other economic theory problems of which one has to be aware when one tries to operationalise sustainability. We should like to point out that these chapters consist of an in-depth discussion of selected themes from Chapter 3, and may therefore appear a little technical.

4
Sustainability from an Economic Perspective*

4.1 The preconditions for an economic analysis of sustainability

Our modern age offers enormous opportunities now and in the future, at the same time as we have allowed ourselves the potential of causing enormous destruction. It is becoming ever more relevant – both politically and scientifically – to seek an answer to the question of how our actions today limit our options and freedom of action in the future.

Sustainability as justice for all generations is an important theme, precisely because our generation – using the gifts of modern technology – is able to deplete vital natural resources and inflict serious long-term harm on significant environmental resources. Three questions immediately spring to mind:

1. What obligations does this generation have with respect to future generations?
2. How can we decide whether we are meeting our obligations?
3. How can we ensure that we meet our obligations?

Our analysis will build upon economic theory and has an anthropocentric basis: Resources must be managed if they are to be able meet the needs of both the present and future generations. Because the interests of future generations are also included, the conclusion that biodiversity should be maintained will be in line with economic

---

*This chapter builds mainly on Asheim (1994c, 1995a).

theory. Even in circumstances where this would not show a register-able benefit for the present generation, it may well prove to be of benefit for generations to come.

Economic theory is not limited to the analysis of the benefits and resources which are on offer in the market place; economic theory is not concerned only with what is commonly called 'narrow economic interests'. The existence of a free market system is, however, significant: market prices can provide information which can support good collective decisions, and a market economy is an economic system which, in certain circumstances, can provide good environmental management.

The good qualities of a market system come from the individual responsibility borne by each individual. If an individual wants to consume a certain commodity, or a company wants to use a certain good as an input in its production, then the individual or company must go to the market place to buy it. The payment received by the seller allows her the opportunity to utilise a bundle of resources valued at the same amount as the goods she sold. In this way, the purchaser becomes responsible for compensating the seller for the value represented by the good which was sold. Under ideal con-ditions, this system leads to goods and services being used where their value is greatest.

Even under ideal circumstances for the use of resources, however, the free market system has one significant weakness: it does not ensure a fair distribution of income. Individuals who already, for one reason or another, are lucky enough to receive a good income will also be able to enjoy a larger part of the overall goods and services. Especially when such differences in income levels are not a result of differences in the efforts expended in earning them, then it would seem rather less than fair and give rise to redistribution of income by the authorities. In fact, a market system has in practice the inherent weakness that many goods, including many environmental resources, are not for sale in the market place. This means that individuals and companies can cause the availability of environmental resources to decline through pollution, without being made responsible by having to compensate those who have suffered the loss of these resources. Economists call these negative external effects, and the authorities, in turn, often attempt to apply environmental policies to make the individuals or companies responsible for their pollution.

In the sustainability context, this means that even under ideal con-ditions a free market economy will not necessarily ensure justice down the generations. An important complicating factor is also that the authorities are, of course, representatives of our generation. Transfer from our own generation to future generations would necessarily assume the character of charity. The existence of external effects which inflict damage on environmental resources is also significant for the question of sustainability. Such environmental harm will often have long-term effects, so that the consequences are just as serious for future generations as for our own. In such cases, the extent to which the authorities are willing to make the polluters responsible for their activities will depend on the extent to which our generation wishes (and is able to) to transfer the income to the generations to come.

We shall now prove that, even under a great deal of exceptionally simplified preconditions, sustainable development will be a goal which it is difficult to plot a course to and steer the helm towards. For a more detailed explanation of these preconditions we refer to Asheim (1995a). Here is a brief review of them:

(a) The living standards of any generation depend upon on the heritage it receives from previous generations and which it hands down to succeeding generations. For example, a generation which consumes fewer natural resources, causes little deterioration of the quality of environmental resources and accumulates a lot of human and man-made capital, will have a lower standard of living than it would have had if measures which benefit future generations had not been applied. The expression 'standards of living' is assumed to include more than material consumption: it includes everything which affects the human situation, including the quality of life provided by access to environmental resources.

(b) The basis has been simplified to discussing justice to, and redistribution between, the generations. For this reason, we ignore the fact that a more egalitarian distribution within one generation and between nations will provide higher living standards than an unequal distribution of welfare.

(c) We will ignore the problems involved in measuring standards of living.

(d) Nor will we, explicitly, take into account the fact that the heritage which we hand down to future generations will exhibit uncertain consequences – for example, the accumulation of environmental toxins and reduction of biodiversity.

(e) In addition, we will not explicitly discuss how population growth threatens to undermine the possibility of attaining sustainable development. In order that this analysis should be easier to understand, we will assume that population growth is equal to zero.

4.2 Our obligations to future generations

The most common definition of sustainable development as suggested by economists runs like this: *Our environmental management is sustainable if the living standards which we gain for ourselves can also be enjoyed by future generations.* If this demand is also directed at our descendants, then the redefinition should be as follows: *Our resource management is sustainable if it represents the first part of a possible trend of non-declining living standards.* This definition does not mean that the state of development which is actually reached will have non-declining standards. The reason is that sustainability does not exclude the possibility that one generation can sacrifice itself for future generations to such an extent that it achieves living standards which are lower than previous generations, although all of the generations have behaved in accordance with sustainability. It is inherent in this definition that each generation manages its national assets, that this yields a positive return, and that only the return on the assets is consumed to procure its own living standards. These national assets consist of both renewable and non-renewable resources, man-made real capital, the quality of the environment we live in, peoples' knowledge and experience, the organisation of society and the foreign debts and claims of the nation. Most of this is difficult to measure meaningfully in monetary terms, but that is another question.

Sustainability, as defined above, seems to mean that our generation can agree to sacrifice itself for future generations, while the equivalent agreement cannot be received from our descendants. This asymmetric attitude has its basis in the belief that if we sacrifice ourselves then our descendants will gain more than our sacrifice. To put it another way: our sacrifice is an investment which yields a net positive return. If it means that we are left with lower living standards that future generations can benefit from, this will lead to increased inequality. However, the 'cake' which is available for distribution will increase. In the opposite case, if we just do what we like and leave the necessary sacrifice to our descendants, then that sacrifice would have to be greater than the increase in standards of living which we could benefit from. If our standard of living is higher than theirs, then such policies on our part lead to greater inequality, at the same time as the 'cake' available for distribution will become smaller. In other words, while the redistribution from generations to come to our present generation can be regarded as a positive cost, the redistribution from our generation to our descendants is understood to be a negative cost, because you receive more than the size of the original sacrifice.

According to this view, the belief that investment yields positive returns gives a normal basis for sustainability. We see this as a belief that the Earth is productive. In a geological perspective, there are millions of generations which could succeed us. We would not, therefore, be making a great mistake if we were to model this as if we were the first of an infinite succession of generations. It is uncontroversial to claim that we, and our highly developed technology, have the opportunity of destroying the resource base to such an extent that all future generations will suffer reduced standards of living, or even have their very existence threatened. The opposite is also true: if we make decisions today which will benefit future generations, it will increase living standards for all future generations. If we believe that this postulate is viable, then it is not difficult to admit that our generation is obliged not to avail itself of living standards which cannot also be shared by future generations. In this case, our final sacrifice will produce an infinite increase in future standards of living and, at the same time, reduce inequality between the generations.

A rule for sustainability could be based in comparison of levels of living standards, in this way maximising the living standards of the generation which has the lowest standards (Rawls, 1971). In general, such a welfare function will lead to all generations enjoying the same standards of living: the result would be complete equality. The level of living standards which each generation ensures for itself when following such a course can be regarded as the maximum level of sustainability – that is, the level which satisfies the condition that future generations should be able to enjoy an equivalent level of living standards.

It should usually be possible to find weightings on each generation's utility which would be independent of the form of the utility function and give an egalitarian course of development. These weightings would depend on manufacturing opportunities and, as long as there is positive productivity, place greater emphasis on the utility of past generations. This means that future utility will be discounted. Because the result is completely egalitarian, there will be no reason to claim that such utility discounting is unfair to the generations to come.

It is our basic premise that the expression 'sustainable development' does not require that a definite welfare function is chosen as a normative basis. The meaning of the expression is not that one certain line of development is sustainable and therefore better than other lines. The important fact about the expression is that it points out that certain possible trends are non-sustainable and are therefore not acceptable social options. In practical political terms this demands that rules for resource management are drawn up which provide information about whether subsequent developments are sustainable or not, and that they indicate how collective action can prevent possible non-sustainable resource management.

5
Operationalisation of Sustainable Development: Economic Principles for Resource Management[*]

5.1   The demand for strong sustainability

By assuming that each generation's living conditions are dependent upon the heritage it received from previous generations, and what it hands down to generations to come, and not least that this relationship is known, it may seem to be an easy task to decide whether our generation's resource management is sustainable or not. If the heritage we received, and which we are going to pass on, consists of only one capital good the principle for deciding sustainability will be as follows: *Let the stocks of the capital good which is passed on be at least as great as the stocks which were handed down (that is, the national wealth remains intact).*

If sustainable development is a stationary process, the requirement for resource management which is placed before every succeeding generation would be that they must leave behind at least as much as they received of each capital good. This sort of demand is often linked to the expression 'strong sustainability' (see, for example, Daly, 1992). It should not be understood as a definition of 'sustainability', but as a principle for how sustainability could be achieved. It is often presented as a principle which must be followed because the accumulation of one capital good (for example, knowledge) cannot compensate for reduced access to another (for example, natural and environmental resources). This pessimistic attitude to the opportunity for substitution must, however, be taken in conjunction with the optimistic view that it is feasible to realise acceptable living conditions for the global population by following a stationary process.

---

[*] This chapter is based mainly on Asheim (1995a, 1995b).

Strong sustainabiity means stringent demands for management. As a result, no generation would be able to exploit non-renewable natural resources, even under the condition that the resources are of no value while they are unexploited stocks, and do not harm the environment when they are exploited. In practice, however, the proponents of strong sustainability promote the view that it will be possible to exploit such resources as long as they are compensated for by accumulation of other renewable resources. The stocks of these renewable natural resources must be used in the same way as the non-renewable resources they replace (see, for example, the discussion of ecological premises in Section 9.1).

5.2   The demand for weak sustainability

There may be reasons for arguing that sustainable development will not be a stationary process. Human activity will lead to depletion and deterioration of the quality of natural resources. For this reason, if a generation is to meet its obligations to future generations it must replace such loss of natural capital by investing in human capital (especially knowledge), better social organisation and man-made capital. It is important that such compensation, in the form of knowledge and man-made capital, is of such a nature that it does not require further depletion of natural resources and deterioration of environmental resources for it to be exploited. In other words, knowledge must produce a 'sustainable technology' upon which man-made capital can be based. One example of a technology which is not sustainable is a technology for the extraction and exploitation of non-renewable resources through a process which inflicts long-term damage on environmental resources. On the contrary, technology which harvests renewable resources in a maintainable manner without causing pollution could be characterised as sustainable.

The principle that every succeeding generation must compensate reduced natural capital through accumulation of human and man-made capital is often known as the demand for weak sustainability (see, for example, Pearce and Atkinson, 1993, p. 104). Not only is this a definition of sustainability, it is also a principle for achieving sustainability. The supporters of the demand for weak sustainability may be said to have faith in the possibility of human and man-made capital replacing natural capital within certain frameworks. This may well be combined with pessimism concerning the possibility of achieving acceptable living conditions for the population of the world by following a stationary course.

If a principle about weak sustainability is to be applied, it must be possible to assess whether the reduction to which stocks of natural capital are exposed is actually compensated for by increases in stocks of human and man-made capital. If this is to be possible, there must be relative prices for the various capital stocks. Let us assume that there are market prices for all capital goods, including natural and environmental resources and intellectual capital (for more about the problems involved in such valuation, see Chapters 3 and 7). Let us assume that central government environmental policy holds polluters responsible for negative external effects.

Under these conditions it is usual for economic literature (Mäler, 1991, p. 11, Hulten, 1992, p. 17; see also Solow, 1993) to claim the following: *The resource management of a generation is sustainable if, and only if, the increase in human and man-made capital is at least as great as the reduction in natural capital.* Even under the most ideal conditions, however, this will not be the case, because it is entirely possible that the market value of our generation's increase in human and man-made capital exceeds the reduction of natural capital, at the same time as we enjoy living conditions which cannot be shared by all future generations (Asheim, 1994a).

Under such ideal conditions the result achieved will be as follows: *If development is egalitarian – so that each generation ensures itself the maximum sustainable level of living conditions – then it would apply to each generation that the increase in market value of human and man-made capital is exactly as great as the reduction in natural capital.* This is known as Hartwick's principle (Hartwick, 1977; Dixit, Hammond and Hoel, 1980).

The above paragraph does not mean that there is no principle which can decide whether our generation is acting in accordance with sustainability; on the contrary, it is a result which characterises a certain course of sustainable development where there is a completely egalitarian distribution of living conditions between generations. Even along such an egalitarian course, the accumulated market value of the capital stocks is constant. This is because the individual capital stocks (especially the remaining stocks of natural resources) will be able to increase in market value over time. These sorts of capital gains reflect, on the global level, reduced productivity and for this reason cannot be regarded as a source of a maintainable income (Asheim, 1994b). A principle of non-declining market value of the total capital stocks is therefore, even under ideal conditions, an non-viable criterion for sustainability.

5.3   There are no simple principles for sustainable resource management

Under less ideal, although more realistic conditions, we shall see that the possibility of constructing principles for sustainability are reduced even further:

1. *For many capital stocks there are no market prices.* This applies especially to human capital (knowledge) and many forms of environmental resources. Attempts at calculating prices for such stocks for this purpose are very uncertain and must be regarded with great scepticism (see Chapters 3 and 7 for a more detailed discussion).
2. *The population of the world is far from constant, it is still increasing at a great pace.* This means that each generation will leave more than it inherited. The size of such an accumulation is a complicated equation; it is partly dependent on the size of future population growth.
3. *Policies for sustainable development will principally have to be exercised on a national level.* For this reason, it is important to draw up principles for sustainability which each individual country can adhere to. If any country is basically more wealthy than any other (and it is, after all, so) following principles for sustainability would not necessarily lead to international parity.

It can be proved (Asheim, 1994b; Asheim and Brekke, 1994) that a sustainability principle for an individual country does not only depend on the degree to which it replaces reduction of natural capital through increase of stocks human and man-made capital. It will also depend on the nature of the capital stocks the country possesses. For example, a country which possesses large stocks of natural capital with a value that increases over time will have proportionally less responsibility for ensuring sustainability than a country without such stocks.

For Norway, a small, open and resourceful economy, it is interesting to examine the sort of management principles that this sustainability analysis leads to. In simplified models of such countries (Sefton and Weale, 1994) it has been proved that compensation for reductions in stocks of natural resources must be paid for by the country which consumes the resources, not the country which produces them. If a national expression is to include this, it must take into account capital gains, but this does not usually happen.

The empirical analysis of sustainability which Pearce and Atkinson (1993) present provides an illustration of this problem, which is also of interest from the point of view of Norwegian sustainability. Among other things, this refers to the huge rate of increase in man-made stocks in Japan and Indonesia. Japan (which has restricted stocks of natural capital) reduces its stocks of natural capital to only a minor degree so that the country appears to be very sustainable. Indonesia (with its huge stocks of natural capital) reduces its stocks of natural capital to a very great extent so that the country appears as marginally non-sustainable. If, on the other hand, sustainability within each country means that the countries must reinvest natural resources in relation to its consumption of resources, then greater demands will be placed on Japan and fewer on Indonesia. It may even be that Indonesia will appear as a more sustainable country than Japan.

So what possibilities remain to translate the expression 'sustainable development' into practical policies? Such principles must require indicators for accessibility to renewable natural and environmental resources which are essential for production of goods which satisfy basic human needs. Such primary goods include food, water and air. Indicators which show the ability of the Earth to develop a maintainable production of food and water will, therefore, be valuable (see, for example, Hansen, 1993a, Part 1). A complicated condition in this connection is that, in practice, food is produced using both renewable and non-renewable resources (fossil fuels; both directly and to manufacture artificial fertilisers as well as pesticides). These are part of the food chain and contribute to the reduction in accessibility of natural and environmental resources. In the same way, indicators which show accumulation of greenhouse gases (and therefore the potential for global climatic change) will be important. Pragmatic principles can be formulated on the basis of such indicators.

We will return to this, a central feature in this study, because we have recently come to understand that, in the field of economic theory, it is necessary to gather scientific data and environmental expertise on trends in important indicators which can act as 'warning lights' for the environment and the resources which we are interested in managing, based on the demand that life-supporting resources and ecosystems must be preserved for our descendants. A central feature for operationalisation of sustainability will, therefore, be to find a set of premises for sustainable development and design these so that they can be aligned with the economic planning tools used by central government in its long-term planning.

Because these sorts of indicators will have to be based on physical dimensions, this approach will be reminiscent of the sustainability principles which the supporters of strong sustainability have promoted (see, for example, Daly, 1992, p. 251). Our argument for presenting such principles is, however, not a lack of belief that increases in human and man-made capital will be able to replace certain stocks of natural capital; from our point of view, the problem is that it may be difficult to decide the size of the compensatory investments. This difficulty is partly connected with the problems in deciding what 'compensatory investments' means, and partly with basic problems in valuing the environment. This will be the theme of Chapters 6–8.

6
Compensatory Investment[*]

6.1   What is compensatory investment?

Business and industry are finding it to be strategically more and more important, and therefore more popular, to exhibit social awareness and concern for future generations in their resource management. Measures they are implementing include voluntary environmental audits within a system of annual environmental accounts. In the case of new invest-ment which may pollute the environment, companies may show responsibility in various ways. One is by setting aside funds for envir-onmental measures which are intended to compensate for the damage caused by the original investment. The best known of these compen-satory business measures is the replanting of forests in the Central Americas by companies operating coal-fired power stations in the USA, who market reforestation as a measure which absorbs as much CO_2 as is emitted by the increase in coal-fired power stations in the USA. The reforestation can be regarded as the price which the company pays to protect itself against a poor environmental reputation, as this could have a negative effect on its opportunity to win new concessions for future expansion. Measures of this type are known as 'compensatory investment' (CI), and have been the subject of a great deal of attention in connection with attempts to operationalise sustainable develop-ment.

The literature about CI appears as an attempt to link theoretical, economic definitions of sustainability through generations with the practical demands of sustainable policies. CI is an attempt to avoid some of the difficulties of macro-economic planning, and at the same

---

[*] This chapter is based on Farmer (1995).

time avoid unnecessarily complicated solutions by setting simple goals, indicators and guidelines which, as a whole, are supposed to protect against a further erosion of future welfare standards. CI is, in other words, presented as a practical and useful proposal for a policy of sustainable development.

CI measures simply – year after year, project after project – protect the capacity to maintain welfare by transferring resource stocks to the future. Assume now that our basic economic activity results in a decrease in the value of our resource stock which is transferred to the future. In such a case, the CI plan must contain special measures so that sufficient savings are achieved to maintain the level of the actual resource stock. This enables us to avoid having to go through complicated general equilibrium analyses of macro-systems when we want to say something about the sustainability of investment programmes and other economic policies.

6.2   Illustration through 'measure-for-measure' compensation

'Measure for measure' CI proposals refer to individual projects in the economy. They portray a situation as sustainable if investment in the protection of another resource prevents depletion of the relevant resource from resulting in a reduction of welfare. Such measures can be further divided into two sub-groups:

1. replacements of the same type, and
2. replacements with the same economic value (value replacements).

'Replacement of the same type' can be illustrated by activities such as the chopping down of a forest, and compensating for it by protecting another forest which would otherwise have been cut. This probably most closely corresponds to what we have already defined as 'strong sustainability' (Chapter 5). By 'value replacement', on the other hand, we mean chopping down a certain forest, and compensating for it by investment in the protection of certain resource stocks of a corresponding value. This more or less corresponds to what we have already defined as 'weak sustainability' (Chapter 5).

The appeal of these CI rules is their simplicity, sturdiness and clarity. CI demands that decision-makers solve each detail by a series of complicated measures, as sophisticated planning is required to guarantee the desired results. If we assume a precise 'measure-for-measure' strategy by paying for things as we use them, then we can give a virtually 100 per cent guarantee that we will leave as much to descendants as we have today. This logic appears to be cast iron: in principle, if everything that is used is replaced, then it will be possible to guarantee that there will not be any net reduction of the value of resource reserves.

There is, however, a serious problem in cases where the unit value increase is the result of increased depletion of resources which, in its turn, means that there will be physically less and less left for each of us (see the discussion of weak sustainability in Section 5.2).

These CI rules are also completely dependent on our ability to identify the opportunity cost of the resource to be exploited and the resource to be preserved. It especially demands independence between the conditions which lead to use of the resource stock we compensate for and those resources which are preserved.

These aforementioned 'measure-for-measure' CI rules must be more closely defined, so that if we fell a forest then we must preserve a forest that was originally intended for felling. There would be little value in protecting a forest which would have been preserved anyway. Not only that, there would be little value from a CI perspective if the forest owners – after having felled one forest and preserved another – decided to fell a third forest which would have been allowed to grow if our CI measures had not been implemented. In other words, the CI rules must guarantee an unequivocal increase in the net preservation of the resource reserves which are actually passed on to the future, through the measures which are implemented. Our savings measures must be isolated and not cause any consequential reactions anywhere else. The danger of the complexity of sustainable policies being camouflaged behind a rhetorical veil of simple implementation must, however, be avoided.

6.3   Compensatory investments – a blind alley

The basis of the CI concept is the formulation of solutions to sustainability problems, based on the view that a holistic understanding does not manage to create the desired results. The problems are, however, made visible through the complex mechanisms and mutual dependencies of the entire system – for example, in the shape of expected future shortages of a certain resource.

At first sight, the introduction of CI appears to be tantalisingly simple. The sustainability problem is simplified down to a few indicators which can be easily manipulated marginally, but which cannot be separated from the rest of the system. For this reason, we have initially argued for building the compensatory investment proposals on a set of preconditions about a general equilibrium where it is possible to identify, and measure the loss which results from too little savings. Next, however, we reject these preconditions so that we can promote our partial equilibrium solution, in other words, the so-called 'CI plan'.

In this way we are far from helpless when faced with the sustainibility question. In fact, the simple CI proposals are not sufficient in themselves, and viable solutions must be developed in the same context, and under the same set of preconditions that define the problem. In contrast to what we have been led to believe, compensatory investments do not give us a simple, user-friendly and robust measure with which to achieve sustainability goals. This is in agreement with the conclusion we arrived at in Chapter 3, where we examined the feasibility and desirability of simplifying the operationalisation of sustainability by 'greening' GDP.

6.4   Sustainability when there are gaps in our knowledge

Thus far, this discussion confirms the gloomy prospects for a comprehensive strategic management of sustainability from the analysis in Chapters 4 and 5. This must, however, not be understood as an encouragement to throw in the towel. On the contrary, it becomes even more important to be able to give signals about the correct direction of development, and a theoretically sound basis for analysis, when it is not possible to design simple measures because of limitations in technology. This leads us to a better understanding of the problem and a better foundation for informed choices and the setting of priorities. In the absence of perfect theories and exact estimates of how the economy and environment will react to our measures, it is all the more important to exploit the knowledge we do, in fact, have.

We know at least enough to focus on sustainability in the context of policy design. We have discovered that it is not that easy to purge the domestic product of all that we do not like, in order to be able to produce from it an operational and useful welfare goal (see Chapter 3). We also know, from our previous analysis, that even if we have solved this problem, the problem of compensation for a savings deficit by choosing investment will be considerable and require major theoretical and empirical clarifications.

On the other hand, we also know that our valuation of what we have at present shows how important each good is to us, compared with the other goods and services we consume. You would hardly receive long odds if you wagered that our descendants would also appreciate a clean environment, good schools, beautiful countryside, good health, healthy food and safe housing in a way which does not differ much from our own priorities. If so, then it follows that a measure which protects against depreciation of the basic needs resources gives a good *a priori* basis to protect the welfare of future generations – in other words, the protection of sustainability.

Price information must be exploited with all of the conditions and weaknesses which can be proved. Prices provide extremely important signals as to what people really appreciate and can, therefore, contribute to the clarification of the direction of welfare trends when price and quantity are combined for the resource stocks in question. In this way, a sustainability strategy can be based on quantifiable net changes in the actual stocks of the resources in question, and by multiplying them by present prices (at least for those resources which are for sale). If an increase in value is noted, this could indicate that the measure strengthens sustainability, and the opposite if the calculation indicates a loss of value. However, as we saw in Chapter 5, it does not have to be this way. For petroleum resources, where price forecasts can vary strongly from one year to the next, major problems arise when fixing value trends (see Chapter 3). For common resources which have no market value, it is, of course, far more difficult to assess how much the value of the resources has changed. There are, however, various evaluation techniques (see Magnussen, 1994; Magnussen and Navrud, 1995) which can indicate rough intervals, and often the movement of the trend – in other words, if the trend seems to indicate sustainability or not. No more certain answer than this be given because of the sustainability conditions we proved to be necessary in Section 5.2 concerning weak sustainability.

The example with the uncertainty of the petroleum resources in Chapter 3 is just one that illustrates that detailed resource accounting is not sufficient for sustainable resource management. The balance between drawing on a resource today, or in the future, is very complicated. If a resource is drawn on today, and used immediately, it will not be available for later use. We shall illustrate the many, and different, problems this raises in the sustainability connection by looking at the decision between forests and wooden building materials (Box 6.1).

Box 6.1   Opportunity costs of a tree

If a tree is felled and is used to build a house, we lose this volume of wood *plus* any additional volume from the tree's growth which could be used as future building material. (To simplify matters, we shall ignore any other pleasure we could derive from leaving the tree standing.) Looked at this way, the future has been deprived. We can look at the opportunity cost for this tree when it is felled today as the price of an even larger building tomorrow, since the price has been discounted by means of a discount rate. We will have to be satisfied by a smaller building today, instead of a larger building tomorrow, because we are impatient and want to consume now. This discount rate can, therefore, reflect our impatience or short-sighted perspective, but it can also have other explanations.

By felling the tree today, and building a smaller house, we are declining to use a larger tree tomorrow which could have given us a larger building. It could, however, also be argued that we could use the building we have built in order to produce and consume things that we would not otherwise have produced or consumed. It is very possible that the product manufactured in this building today will be of benefit to posterity. These are all complicating factors in the calculation, as chopping the tree down now may actually benefit the future, even though we have to turn our back on a growing tree and its other beneficial effects.

The discount rate is therefore an expression of, for example, the building capital's opportunity cost for the builder who chooses to acquire a building today, instead of waiting until tomorrow. His determination to pay extra to get the smaller building today is equivalent to what he expects to earn by producing those things in the building today that he would not have been able to produce without the building. If this was a widespread preference in the market, it would push up both the prices of building today and the price of wood-based construction materials. How would this affect posterity? The future would inherit fewer trees and fewer buildings, leading to an immediate price to pay in the form of reduced resources. At the same time, posterity would inherit a capacity for production of

> **Box 6.1** *continued*
>
> the range of objects which could be produced in these buildings, things which would obviously be widely appreciated if manufacture of them in these buildings was found to be profitable. As everything is dependent on everything else, the trees should be preserved, because the anticipated wealth generated by sale of the objects in the future will increase future demand for large houses, and to build larger houses, more and more trees must be felled.

In this conceptual example, felling the tree does not leave posterity any the worse off. Instead, we must weigh the alternatives. The probability of the welfare of posterity being poorer if it inherits only trees which are potential building material, but no manufactured objects, is high. In the same way, posterity would no doubt be extremely ungrateful if it inherited only wooden buildings, manufactured objects and no trees.

6.5 Safe minimum standards (SMS) to achieve sustainability

We have established that both improvement and degradation of welfare levels of future generations is possible in an efficient economy. A renewable biological resource which is necessary for the survival of humanity can exhaust itself in an economy where all markets function perfectly (Farmer, 1995). The quest for the most efficient solution in large, system-wide optimisation programmes risks exposing society to catastrophic reactions within the limits set by statistical confidence ranges. Against a background like this, it has been suggested (Daly, 1992) that measures should be put in place to achieve a 'reasonable' solution which is stable and institutionally feasible. This should secure society against possible undermining of future welfare opportunities. We shall take a closer look at insurance plans like this to secure sustainability.

We all agree that it is extremely difficult to identify and calculate what a sustainable development is, whether locally or globally. In addition, there are the very difficult, and often underestimated, problems caused by changing to policies which enable developments to follow such a course (Haavelmo and Hansen, 1992). Given calculation problems like this, it will be particularly important to be able to register trends in important resource stocks. A time can come when one or more special biological resources can be threatened with extinction, making a special focus on trends in the physical size of a certain resource necessary if we are to be able to implement countermeasures when the risk of it sinking below a specific extinction threshold arises. This critical threshold is, of course, difficult to determine for any single resource, but it is important to ensure that the level of vital resources does not end up below it.

We need knowledge both about growth patterns of natural resources and of the economic dynamics of use of them during the biologically determined regeneration period. We must determine the size of a safe minimum stock of the resource which will be necessary to maintain the level of welfare, although this entails first confirming that the present size of the resource allows a safety margin which is large enough to permit us to both reap the benefits of the resource and to maintain a sufficient reserve to allow for uncertainty as to the viability of the resource. In many cases, this safety margin can prove to be extremely large (see Hansen, 1993a, Chapter 11; Farmer, 1995), and from this we derive the expression 'safe minimum standards' (SMS).

Such standards are recommended as sensible and practical measures which support any sustainability programme. If the knowledge required to implement a central sustainability programme is lacking, the significance of securing such minimum levels of welfare-supporting resources will assume increasing value, as insurance against the undermining of sustainability – or, in the worst-case scenario, against survival. It is not, however, just a lack of scientific knowledge which stands in the way of development and implementation of a sustainable development plan. The political situation and processes will rarely be in place which enable collective integrated policies in societies with open political structures. On the contrary, when there is wide agreement about the necessity of securing such resources, for the sake of the welfare of the country and its future, such structures will create a basis for approving their SMS.

SMS are no replacement for central sustainability strategies, but are rather, a critical alarm which must sound before an irreversible and destructive process of development is reached, as a result either of political inaction or an undetected error in a previously existing sustainability programme. Efficient interplay between economic and ecological management of the limited resources of society is decisive for whether such a programme is feasible. It would obviously be easier to gain acceptance for such SMS measures if the society has an efficient economy, generating profits which the community can devote to securing the common future. A better understanding of the interaction between economy and ecology is, however, also vital because it allows decision-makers to develop institutions which exploit precisely this interplay in resource management.

In this way, the number of areas where it would be necessary to implement SMS, in the absence of sufficient knowledge about the resistance and survivability of the resource within the existing economic and institutional system, will be reduced. This leads to significant cost savings, and greater freedom of action, compared with a society where one is forced to introduce stringent SMS in a wide range of areas because of a lack of ecological and economic knowledge and ability to cope. Such an efficient society will contribute to the preservation of greater parts of the resource base for posterity. In the opposite case, the need for, and the costs involved in, introducing SMS in many areas could develop into the greatest obstruction to approval of the necessary SMS, and therefore, in itself, represent a threat to the ability of the society to survive. The middle road must be sought for the introduction of SMS, as both too many and too few minimum standards will be a threat to survival.

7
Environmental Valuation as a Basis for Environmental and Resource Management[*]

7.1 Using environmental pricing

In principle, the environment is a free gift. Mountain hikes are free, breathing is free and car exhausts are free, even when they pollute the air everyone else breathes. Environmental benefits are common benefits, everyone has the right to enjoy them, and this is precisely the basis of many of our problems with our natural resources and the environment.

In Chapter 3 we presented the valuation of the environment in monetary terms as a chronic headache. There are, however, arguments for putting prices on environmental effects so that they can be included in benefit–cost analyses (BCA). One of these is the need for a tool which can improve the ability of decision-makers in the public and private sectors to compare directly the benefit and cost effects of road, rail and airport development, petroleum activities, hydropower projects, industrial establishment, facilities for the storage and treatment of special waste and other projects which are covered by the impact survey rules in the Norwegian building regulations[1] (and certain projects which are not covered by these regulations, but which also have positive or negative environmental effects).

Another related argument is that environmental effects, unless they are explicitly priced, will be implicitly priced when decisions are made. Just imagine the situation that a hydropower project will yield a net annual mean benefit (net income) in the vicinity of NOK 35 million without taking into account the environmental costs, and

---

* This chapter is a modified summary of Magnussen (1994) and Magnussen and Navrud (1995).

that the decision is made to go ahead with the project. This means that the decision-makers have implicitly valued the environmental costs as less than NOK 35 million per annum. If we presume that the project and the environmental effects have an impact on the welfare of all of Norway's 1.77 million households, then it means that each household is willing to pay less than around NOK 20 per annum to avoid the potential environmental effects of the project. As decision-makers are often unaware that they are making this implicit valuation of the environment, the result can often be strongly varying and inconsistent prices for the same environmental effect. This will, in its turn, lead to inefficient socio–economic management of our resources, including natural resources and environmental quality. It must, then, be better to use existing methods for valuation of environmental benefits to arrive at explicit and more consistent environmental prices.

Environmental valuation is a possible contribution to our understanding, and eventual addressing of, these problems. The valuation of environmental benefits is built on the theory that nature has a value in human terms. The methods of valuation attempt to measure all, or parts, of the total socio–economic value of changes in the size or quality of environmental benefits. This includes both the value of use and non-use, where non-use can be regarded as people's use of the intrinsic value of nature.

Economic analysis of various forms of market failure or market imperfections are relevant for many problems for natural resources or the environment. This chapter will take a closer look at why there are weaknesses in the market for environmental and natural resources, and we will discuss the conditions under which environmental valuation can be an aid to the correction of these market weaknesses.

7.2 Market imperfections, policy failure and public goods

A much-used expression in economic welfare theory and resource economies is that of 'public goods'. Various definitions and interpretations of this expression have been in use. One common definition states that a pure public good is characterised by the fact that: (1) when it is first available, it is available to everyone (non-exclusive) and (2) one person's consumption of the good does not reduce another person's opportunity to consume the same good (non-rivalling). There are, however, a number of intermediate forms because non-excluding and non-rivalling do not necessarily occur together, and such expressions as 'partly public goods', 'quasi-private goods', 'club goods' and so on, have been used to denote these intermediate forms. Various combinations of rivalry and exclusion have different characteristics with respect to the opportunities for the good to be distributed through the market and the opportunities for Pareto-efficient distribution – that is, it is not possible to improve conditions for some without at least one person's situation deteriorating.

Analysis of the goods which are characterised by non-rivalry and/or non-exclusion are relevant to many of the problems connected to allocation of natural resources – for example, management of fish and game stocks, protection of nature, biodiversity, management of public areas, exploitation of oil and the water table, and the supply of natural resources and environmental goods from corporations in the public sector. In this connection, economic analysis of the ability to deliver services, valuation and rationing between the users can contribute to the identification of possible solutions to correct the waste and lack of efficiency which otherwise arises.

Rivalling, exclusive goods are private goods, such as bread and milk, where the preconditions for efficient production, distribution and consumption are 'according to the text book' . By this, we mean that such goods can be produced and allocated efficiently by a well functioning market, without distribution problems arising, provided the distribution of income is acceptable to society to begin with.

There are, however, many goods, services and resources which are non-exclusive. Non-exclusion arises from a lack of, or weakening of, ownership rights, and results in inefficiency. Without the possibility of excluding someone from the use of a resource it is impossible to demand a price for its use. Under conditions like these, prices are unable to ration either the good or the resource between the users, or to yield income for the production of the good, or maintenance and preservation of the resource.

Because it is impossible to demand payment for rival, non-exclusive goods they cannot be offered for sale in private markets. Such goods can be supplied by private, charitable activities or by the public sector, which can finance them by general income. If it were physically and economically feasible to do so, these benefits could be the subject of exclusion in the market-place, allowing the public sector to offer them and demand a charge for their use. If exclusive and unweakened ownership rights were specified and protected, then the private sector could offer them in a socially efficient way.

For a society which values private initiatives, the solution to this problem seems clear: establish unambiguous and clear ownership rights for the good. This would allow independent, self-reliant individuals, or homogeneous groups, to ensure efficient results. Many societies have specified exclusive ownership rights for previously non-exclusive goods and resources. One example is overgrazing of common land. When many societies conferred private ownership on land which had previously had been open to all users, the overgrazing stopped. It should, however, be pointed out that many other sophisticated common rights regimes – for example, in Africa – have also proved themselves to be very sustainable until external forces were given 'superior' powers for various reasons (see Hansen, 1993a).

Some goods and resources remain non-exclusive even in societies which place great stock on private ownership and personal initiative. There are two main reasons for this. One is cultural and political. All societies identify some goods and services which they believe should be kept outside the reach of commercialisation. The other reason is that some goods and resources have a special physical character. Some species of bird, animal, fish and marine mammals are extremely mobile. It is impossible to establish and maintain exclusive ownership of a special fish or to individual small pieces of ocean, and then restrict certain fish and mammals to special areas. In the case of this sort of good the cost of specifying, protecting and maintaining exclusion would by far exceed the eventual benefit of the good. Many of the most important natural resources and environmental goods are associated with this group. It includes the air we breathe, the water in our rivers, seas and oceans, migratory game, and birds and species which have little commercial value. It also includes fish in the oceans, seas and rivers which are too large for private ownership, as well as oil, natural gas and ground water reservoirs which are located under areas which are owned by many people or are a no-man's-land. There are also negative goods and negatively valued resources where efficient exclusion is impossible. This is the case, for example, for plagues, disease-transmitting organisms, as well as, air- and water-borne pollution.

The typical distributive result of non-exclusion is that too little of the good or service is produced compared with the efficient level of supply. On the other hand, this leads to 'too high' levels of negative goods in relation to the efficient level. This leads to overexploitation of resources, which in turn leads to underinvestment in management, preservation and productive capacity of a resource.

There are a number of goods which, when they are produced, are available for all consumers without rivalry. The total output of the good is determined during the production process, but it is not necessary to share the total amount of the goods between the various consumers. Everyone has access to the complete amount of the good, and one person's access does not reduce the amount which is available to others. Although there are not many goods or services which fall into this category, those which do are often important in resource and environmental policy. If the fresh air has sufficient quality, then we can all breathe it and use it as a visual medium without reducing the amount which is actually available to everybody. If the air in a town is polluted, it will simultaneously become poorer for everybody in the town, not just for one individual. An increase of the number of people in an area will not reduce the level of pollution, and in this way it can be regarded as a typical 'public bad'. In the same way, a beautiful garden which is enjoyed by all passers-by would be a public good, because an external effect arises which conveys a benefit upon all who see it. If someone looks at a beautiful view alone, it will not be reduced by 50 per cent if another person comes along to enjoy the same view. If someone derives benefit from the knowledge that a species which was formerly threatened with extinction somewhere in the world has now been saved, then this benefit is not reduced even if others also benefit from the same knowledge.

A number of goods can be enjoyed by many individuals, although up to a certain capacity limit. These are often known as 'congestible goods'. Examples of these include roads, bridges, pipelines, restaurants and concerts. Many natural resources are characterised as congestible goods, including campsites, viewpoints, trails, hunting, fishing and boating areas. Analysis of congestible goods and natural monopolies are, for this reason, relevant for provision of outdoor leisure areas, and their management in this context is discussed further in connection with resource rent taxation in Chapter 22.

Public goods can be treated as a form of external effect. Examples of problems which are often analysed as negative external effects are polluting emissions from industry, various polluting discharges from agriculture, pollution from consumption (for example, vehicle exhausts and tobacco smoke) and any activity which entails noise, aesthetic deterioration or other negative effects for other parties.

It is now generally accepted that when public goods or bads are present the ordinary pricing system is unable to provide an efficient allocation. The fundamental source of this problem is that public goods are non-rival in consumption, which means that one individual's consumption of the good does not reduce its availability for others. Just because one person inhales polluted city air does not change the quality of the air being inhaled by others.

Payment for consumption of such public goods is well known to be inefficient, economically speaking, because the consumption by one individual will not affect other's satisfaction. Price mechanisms can limit an individual's consumption and therefore reduce his or her satisfaction without increasing anyone else's.

7.3 Does use of such values in environmental management encourage more use of environmental taxation and user charges in outdoor leisure activities?

Fees and subsidies for a more effective use of resources

The problem of allocation when external effects introduce imperfection in resources consumption means that there is no normal price which can fit the case, because it is the same (symmetric) for producer and consumer. Economic efficiency requires an asymmetric price mechanism to lead to efficient solutions when external effects are present. This is possible with a so-called 'Pigou tax' (subsidy). Such a charge, or subsidy, means that the authorities set a tax/charge (or pay a subsidy) per unit of the external effect which is equal to the marginal damage (pleasure/benefit) for all victims (or, alternatively, everyone who enjoys the external effects). In this way, they provide the correct incentive for the producers of the external effects, while the consumer meets a price of zero, which is necessary to achieve an effective solution. Figure 7.1 illustrates a situation where the amount X produced, with its associated price P, applies as long as the negative external effect is not taken into account. When the effect of the pollution is internalised into the market, then the authorities introduce a charge $PD^* - PS^*$ which is equal to the marginal social extra cost. This leads to a reduction in the amount produced, and in the demand to X^*. The environmental tax $PD^* - PS^*$ is distributed among the producer, with $P - PS^*$, and the consumer with $PD^* - P$. The less elastic the demand is (the less price-sensitive the demand is) when the angle of the demand curve is given, the greater the consumer's share of this environmental tax will be.

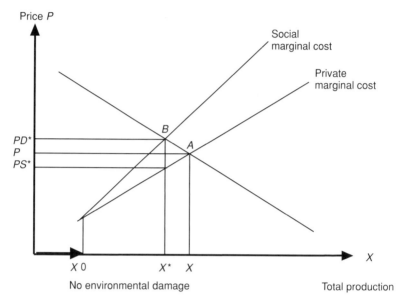

Figure 7.1   Illustration of an optimum environmental tax and its distribution between producer and consumer

The 'many-victims' scenario

Box 7.1   Factory smoke

Let us first examine the well known case of smoke from a factory which pollutes the air in the surrounding district. All of the local population suffer from the pollution, and one individual's 'consumption' of the smoke-filled air does not reduce the amount of pollution for others. The allocation problem here involves two decisions: the factory owner must choose the 'correct' level of pollution-limiting measures to reduce the emissions to their effective level. The 'victims' answer is somewhat more difficult because they can choose between various activities which can protect them against the harmful effects of the external effect. People who live near the factory can invest in air purifiers or move to somewhere further away from the factory. We call such measures 'protective', and these protective measure have no effect on others' exposure to fumes, they are completely private.

Box 7.1 *continued*

The challenge is, therefore, to decide a programme of measures which will result in profit-maximising companies and benefit-maximising individuals adapting themselves in a socially efficient manner. In the case of a competitive market, the solution to this problem is for environmental authorities to set a charge per unit of smoke emission which is equal to the marginal harm to all of the 'victims' (both neighbours and other companies).

This charge will serve to internalise the external costs which the polluting company inflicts on others – that is, these costs will appear in the company's accounts. This forces the company not only to include in its budgeting its normal production costs but also the other sorts of socio–economic costs caused by their production activities. No additional incentives are needed for the 'victims'. It can be formally shown that the harm inflicted on the 'victims' by the negative external effects gives precisely the correct incentives to lead them to take protective measures to a socio–economically efficient extent.

If we look at an example with rival external effects (for example, waste material) we need to examine two possible outcomes. We shall first of all assume that the company has no control over where the waste is deposited. For technical reasons (geographical or meteorological) the waste always ends up in a special place independent of the company's activities in disposing of it. The victim in this case is the owner of the waste-deposit site. The only difference from a case with a non-rivalling external effect is that there is only one 'victim'. For this reason the socio–economic marginal harm is equal to the marginal harm to this individual.

If we assume that the company has a choice as to where it will dump its waste, then the choice of the most efficient spot to dump the waste will be the key question. In this case, the owner of the company must face the different charges which reflect the extent of the harm caused by dumping waste at various sites. When confronted with such a range of charges, any profit-maximising company will choose the site which minimises the harm. A Pigou tax which is equal to the socio–economic harm yields an effective result. As we have also previously seen, the 'victims' can be expected to select the effective level of preventive measures.

For this reason, the conclusion will be that whether or not the externality is rival or non-rival, the correct measure is the one where the charge levied on the person causing the external harm is equal to the marginal socio–economic harm, without any additional incentives for the victims (Baumol and Oates, 1988).

As long as the number of victims is large, compensating the victims will be no effective solution – whether the external effect is rival or not. In cases of transferable external effects – that is, where the victim has the opportunity of transferring the external effect to a third party (for example, where the victim who has waste dumped in his garden can respond by removing it and dumping it on someone else's property) – it is quite clear that there is a need for incentives which lead to effective behaviour by the victims as well as by those who are originally responsible for the external effect.

In this case, a charge must be levied on the victims which is equal to the marginal socio–economic harm which arises for the parties to whom the victim has transferred the effect.

The 'few-victims scenario'

Generally, we talk about situations where many people are involved, these are the most common and important in the politics of environmentalism. In cases where there is only one (or at least very few) producers and one (or only few) victims of the external effects, there would be no need for a Pigou tax or subsidy. On the contrary, in such a case a tax would lead to an erroneous allocation of resources. When both the number of producers and recipients of external effects is small, a system of charges would prove impractical (regulations would have to be drawn up to apply to only a handful of people). When this is the case, the above 'property-rights' solution may well be the most sensible method of controlling the external effect.

In the above analysis, we found that with full information a set of Pigou taxes which is equal to the marginal socio–economic harm will lead to an effective solution in a free market. This result is, however, fundamentally dependent upon a number of other preconditions which raise doubt about the strength of the taxation solution. The extent of uncertainty about the size of the socio–economic harm and purification costs, the extent of regulation of competition in the society and the probability of economies of scale certainly lead to undermining the necessary conditions for socio–economic efficiency. This leads to important reservations about the simplified application of this fascinating taxation theory.

Using price discrimination as a solution

Among other things, the optimal Pareto solution in a market for non-rival goods will demand that each individual must pay a price equal to her marginal valuation of the good. This marginal valuation can, however, vary for every individual. Each consumer should meet a personal price which satisfied the above conditions, in other words, perfect price discrimination is necessary. The problem arises when every individual's personal valuation is to be deduced. If only the individual in question knows her actual valuation, how can others (for example, the producers) know that this individual is not cheating by actually stating a lower valuation? If this person is aware that she will be charged the price that she volunteers, and hopes that others' payment will be sufficient to ensure adequate production of the good in question, then it would be tempting to claim a 'too-low' valuation. If these others paid enough to ensure production of the good, then it would be possible for another to use it without paying, taking on a 'free-rider' role. Experimental research has proved that this would be no widespread problem but the opportunity for various degrees of strategic behaviour can spoil voluntary markets for non-rival goods.

The marginal cost of an additional consumer using the non-rival good when it has been produced is zero. The income of a manufacturer from a non-rival good is the total price paid by all the consumers. If the effective price paid by all the consumers is zero, it is probable that the total income to the manufacturer is insufficient to cover all the manufacturing costs, while higher prices would exclude potential consumers if the valuation of the good is positive but low. This, then, leads to a solution which is not socially effective. There are, therefore, no non-discriminatory pricing systems which will allow achievement of social efficiency in the private sector for production of non-rival goods.

It is possible, under certain conditions, for non-rival goods to be produced in the private sector. If it is possible to exclude the consumers who do not pay, then it is possible for a private manufacturer to receive enough incentive to produce the good. If there is free access to start production of the good, competition between producers can be sufficient to push prices downwards until the total production costs equal the total income, and the rent vanishes. This solution is, however, not efficient because there are consumers with positive, though low, valuation of the good who are excluded from its use. The second-best solution would be to allow the private sector to produce the good and ration it between the consumers without price discrimination and extra profit. Free access to produce non-rival goods is, however, an unrealistic precondition. Price regulation is, in such cases, often used to replace free access to start production in order to prevent the producer from profiting too highly. It is quite clear that this sort of regulation achieves varying degrees of success, but is rarely completely satisfactory.

Non-rival goods can, of course, be produced in the public sector, which can choose to finance provision of these goods from general income and make them available for all consumers. They could, as an alternative and if it is possible, choose to levy a non-discriminatory user price and exclude those who do not pay for it. This latter approach will offer the public sector the opportunity of setting prices such that the income it receives covers the actual costs, or perhaps are higher or lower than the costs involved in production of the good.

The significance of well defined rights

The source of an external effect can, as we have pointed out, often be found in the lack of clearly defined and implemented property rights.

Box 7.2   Access to fish

Take, for example, the case of a lake where all fishermen have free access to fish. The catch of one of these fishermen reduces the expected catch for others, putting an external negative effect in place. The number of fishermen (in socio–economic equilibrium) would be too great because each fisherman inflicts costs on the others in the form of reduced catches, creating a marginal social income which is lower than the value of the marginal product in other activities. In order to correct for this market loss a tax is levied on fishing in the lake which effectively internalises the external costs which one fisherman inflicts upon the other fishermen on the lake. The introduction of the tax will reduce the net gain to the marginal fisherman until it equals his marginal socio–economic product, and this will lead to a reduction to an efficient number of fishermen.

There is, however, another approach which would correct for the imperfection which results from this balance when there is free access. Instead of introducing a fishing tax, presume that the rights to the lake are transferred to an owner (private or public) – for example, in some form of auction. This would then revoke

> ### Box 7.2 *continued*
>
> the fishermen's right of free access to the lake. Presume that the new owner attempts to maximise the profit from the fishing in the lake, and employs fishermen who he or she pays a wage to, and in return owns the catch. This solution which maximises profits means that the lake owner will employ fishermen up to the point where the value of the marginal product is equal to the rate of wages which he or she must pay. The introduction of a clearly defined ownership right will also, in this case be the most socio–economically efficient solution.

A reform of ownership rights can, in many cases, be the preferred measure compared to eliminating the external effect with a tax. Changes in ownership rights are, however, not always easy to implement. There can also be other reasons to desire free access to certain resources. It can, however, be possible to establish certain rights which could simplify regulation of various types of pollution.

When exclusion is possible, specification of the exclusive ownership rights is a political decision. When, however, establishment of exclusive, unweakened ownership rights is not possible, the political choice is more limited. Society often has the opportunity of specifying and upholding laws about who should have access to the good or the resource, and of specifying conditions. Hunters and fishermen may have to buy licences, restriction may restrict hunting and fishing to certain seasons and limits may be placed on the methods used.

In some cases, access rules originate from the social dynamics in a small group which is heavily concerned with the resource, instead of from a central government. These informal access rules may not hold in a court of law, but can still be upheld effectively by the group itself. This sort of voluntary control is becoming increasingly common in the business world, not least as a result of voluntary environmental audits.

Res communis regulations about access to the resource (public ownership/common ownership) do not have all of the effective characteristics of private ownership rights, although they can provide a system of effective rules when exclusive private ownership rights are impossible to introduce. There is great variation in the possible specifications of this type of rules concerning access in the so-called *res communis* field. Take deep-sea fishing as an example. The present level of fishing is not viable in the long term, and is a typical example of the problems which can arise when ownership rights are not exclusive. The costs of establishing exclusive ownership rights in deep-sea fishing would probably be unreasonably high. There is an appreciable and increasing demand for seafood; fish represent a renewable resource which should not be exterminated by inefficient management. There is also a problem in overinvestment in fishing gear and equipment. Many people invest in innovative technology to make the hunt for ever-decreasing fish stocks increasingly efficient. Various forms of subsidy from the governments in several countries have contributed to increasing this sustainability conflict.

In such a situation there are various access rules which can be used. If governmental access rules are to be implemented, then a fishery authority must be established. This body must be able to limit the length of the fishing season. Shorter seasons would probably increase the reproductive fish stocks. If a limited fishing season is imposed and free access is allowed during the season, then the fishermen would compete with each other in buying and using the largest and fastest fishing boats, and the most effective technology for finding and catching fish. The result could easily be overinvestment.

Another approach would be to limit the opportunity to start to work as a fisherman. The probable solution to this would be to licence fishermen and grant licences only to present fishermen and their descendants. Even though this sort of licensing would create valuable property for the fishermen (that is, their licences would gain a capital value) it would do little to protect the fish. The body could choose to limit other fishery factors, perhaps by limiting the size of fishing boats and their capacity to store and freeze fish, or limit – even forbid – use of especially efficient methods for finding or catching fish. This sort of restriction on fishery factors would be relatively inefficient as fish stock protection because the fishermen would quickly be able to substitute other methods to replace those which were sanctioned. This process of substitution would result in an inefficient investment in fishery resources.

The body could try another approach. It could establish market quotas for catches and distribute these quotas among the fishermen. Catches which exceeded the quota could not be sold. Each fisherman could decide the least expensive combination of methods needed to catch his quota. If this quota was transferable (that is, could be sold to other fishermen) the process of dynamic adjustment in the fishing industry would be encouraged. More productive fishermen would expand their operations and technologically advanced companies would enter the industry. Fishing companies which ceased operation would gain additional capital when they sold their catch quotas to new or expanding companies.

This example illustrates some general principles about the general efficiency of various types of rules governing access to a resource. If the problem is overexploitation of a resource, the favoured solution would be to limit the degree of exploitation directly, rather than restricting different sorts of methods of production. When restrictive regulations are necessary, efficiency is gained if the rights are transferable. Finally, it should be emphasised that even the more efficient sorts of *res communis* rules about access would not result in social efficiency. When exclusion is impossible, however, then sensibly drafted rules about access can result in active next-best solutions which maintain productivity in the long term, counteract waste of investment methods, and allow participants to earn sensible and certain incomes.

Environmental valuation and taxation

Valuation of environmental goods need not lead to more extensive use of environmental taxation, although it can make more use of it possible. It can also lead to the level of taxation being set at a socio–economically more 'correct' level. We have seen above that in many cases (of non-rival goods) the introduction of consumer taxation would not provide the socially best solution because it would exclude people who have a positive (although low) benefit of the good, without others gaining increased benefit for that reason. The valuation of environmental goods can, however, also have information value because it reveals how much it is worth for society to maintain this opportunity. The next stage would be for society to determine whether this benefit can defend the costs involved in providing the general public with the good and how they should be paid, if necessary. This also has a more general application. Whatever measures are used, the information about the value of the harm/benefit to the environment would be important for decisions about whether to implement measures, the extent of the harm which is 'correct', and so on. For economic measures, including taxation, the valuation would offer the opportunity to choose the 'correct' level.

Combination of economic measures and environmental standards

There are a number of political measures which have, in their own special way, strengths and weaknesses. An efficient programme of

environmental management must probably use several of these measures in an integrated policy which exploits the strengths (and, if possible, avoids the weaknesses) of the various measures. It is also possible
to implement an environmental policy using economic measures
without going to the length of valuing environmental benefits/harm.
Baumol and Oates (1988) suggest, for example, something that they
think is a practical and effective procedure for environmental
protection – that is, the application of pollution taxation to achieve
predetermined standards of environmental quality.

A certain degree of coincidence in the drawing up of such standards is
unavoidable – for example, the uncertainty surrounding the fixing of
natural tolerances. When such a procedure is suggested, the attempt to
achieve the true social optimum is given up. Apart from the reduced
administration costs, which are made possible by avoiding central directives and direct control, you can show that the suggested procedure, if
drawn up and implemented properly, can under suitable conditions
lead to something approaching minimum costs for society.

An alternative system for achieving predetermined environmental
standards is a system of transferable emission licences. We find that a
properly designed quota system, as with a set of taxes, can also achieve
the standards of minimum social costs. The quota system can, in
certain cases, have some significant political advantages compared with
a system of taxation. It is also presumed to give environmental management more direct control over levels of pollution without necessarily having to introduce a new type of cost on the present sources of
polluting emissions.

Baumol and Oates (1988) also suggest that direct control can be a
useful supplement to a system of taxation for continual maintenance
of acceptable environmental conditions. They base this on the reasoning that rates of taxation are relatively inflexible, and that certain types
of direct control can be easier to introduce, uphold and remove. The
problem is that the environmental quality at any point in time is not
just dependent on the level of emission, but also of such important,
independent effects as wind speeds and rainfall, which determine the
rate of spread of accumulated pollution. As a result of this, we can
expect that coincidental environmental crises can at best be predicted
only immediately before they occur. It would be too expensive for
society to keep the levels of taxation high enough to avoid crisis
situations at any time. It could be less expensive to use direct control
on a temporary basis, despite its static inefficiency. The best result
would be to combine fiscal solutions and direct control to minimise

the anticipated costs to society when certain environmental targets are reached.

Another economic measure is the use of subsidies as a reward for reductions in harmful external effects by those who cause them. In certain, very unrealistic, preconditions, taxes and subsidies set on an equal footing – in other words, give the same effect. Subsidies, used in a more traditional manner are, at least theoretically, a poor replacement for a tax. Even if the two alternatives could be equally effective in reducing emissions from individual companies, subsidies encourage companies to start up in a business, while a tax encourages companies to leave it. Subsidies are also a burden on public funds, and it is more expensive to recover that little extra money than in the private sector. As a result, we can expect that a programme of subsidies would be less effective in reducing pollution than a system of taxation with the same marginal rates. Baumol and Oates (1988) found that during free competition, especially, and if the emissions are proportional to the production levels of the industry and increase monotonously in tune with production, then a programme of subsidies will necessarily yield undesirable effects. Total emissions in a programme of subsidies will, in this case, always be greater than they would have been if no programme had been introduced. Another negative effect of the subsidies is the budget load. Increased subsidies encourage increased public funding, and in many cases the costs to the community of collecting extra taxation can be appreciable, both administratively and because it can increase the negative effects of inefficient price distortions.

Measures to improve the environment can make it more difficult to cope with another important fact; that of the distribution of incomes in society. Baumol and Oates (1988) suggest, on a theoretical basis, that it can be expected that under a range of different conditions the rich can afford, and therefore value the benefit from, a programme for improved environmental quality better than the poor. Programmes which yield the same observable benefit for everyone will probably give a higher welfare gain for the rich, and even programmes where the effects vary by income class cannot be presumed to favour the poor. Nor is it clear that progression in the fiscal system will mean that the rich will bear a proportionately greater share of the costs. Given the type of activities which are the primary candidates for such taxation and the pattern of consumption in various income classes, there is actually reason to suspect the very opposite.

At which level of administration (national/regional/local) should any environmental standards be fixed? Should the central authorities set blanket national standards, or should the local administration set standards which are suitable in their own area? A purely economic perspective would give an unambiguous answer: standards of environmental quality should balance marginal gain against marginal control costs, area by area – standards for pollution which do not cross area limits should also be local. But what if local authorities, in their eagerness to attract new investors and jobs, reduce environmental standards to attract in new companies? Would the result not be ruinous competition between local areas and lead to an undesirable degree of environmental harm? Basically, analyses show that local competition on investment does not necessarily lead to low local levels of environmental quality. Baumol and Oates did, however, find cases where the fears of high levels of pollution when decisions are made locally are well founded. In such cases, centrally decided limits to local choice should be in place.

8
Opportunities for – and Obstacles to – Public Action[*]

8.1 How the authorities can prepare the way

As was shown in Chapter 4, a market economy would not necessarily secure intergenerational justice, even under ideal market conditions. To a certain extent, the problem is that all things that we, as individuals, wish to hand down to our children are not sufficient to ensure that the next generation has as good a start as we had. Under what conditions would there be a basis for public action by governments which are, after all, composed of democratic representatives of our own generation?

First of all, no market system will always make the polluters responsible for the harm they inflict upon others. Environmental policies are a type of public action which aim at forcing polluters to be accountable for the damage they cause. International co-operation on environmental policy will be necessary if we are to confront global environmental problems. International treaties about the emission of CFC (chlorofluorocarbon) gases and CO_2, and protection of biodiversity, are just two examples of such co-operation which would, at the same time, have major consequences for distribution down the generations. It should, however, be emphasised that environmental policies which make polluters responsible for the harm they cause are primarily aimed at preventing waste. In this way, the opportunities to improve the living conditions of one generation, without deterioration in the situations of later generations, are fully exploited. The fact that successful environmental policies manage, in this way, to prevent waste in our management of natural and environmental resources, may not alone prove to be sufficient to ensure sustainable trends. It is quite likely that we are still using so great a part of

---

[*] This chapter builds on Asheim (1994c).

the resources that the possibility for the generations to come to meet their own needs is undermined (see Asheim, 1994c).

The question, therefore, is whether there is a democratic basis for our generation's governments to implement public actions which intend to redistribute incomes from our generation to the generations to come. Again, it seems that the explanation must lie in the existence of external effects, although of another type than those discussed above. It may be that there are two different motivations for us to transfer income to the next generation. One is the concern we have for our own children, leading to an *inheritance motive* within each family. The other is connected to the value it has for us that future generations will have *acceptable living conditions*. However, it matters little for the success of the latter goal exactly who leaves a large inheritance. If a large inheritance is handed down from one family, its cost will necessarily be borne by the older generation within that family. At the same time, there is a gain in the form of improved living conditions for future generations which has a value for everyone who is alive now. This means that there are positive external effects connected with leaving inheritances (Marglin, 1963). This means that governments, made up of democratic representatives of our generation, should wish to implement public actions which intend to transfer income from the present generation to later generations.

How can such transfers be realised as a result of democratic governmental public action? Given the fundamental problem that families, in their decisions concerning their heritage, do not take into account the significance of this heritage on the rest of the community, the most direct measure would be to increase the inheritance motive within each family. The reinforcement of the inheritance motive would also ensure that those who inherit will also pass it on to their own descendants. It is, however, unclear exactly how governments will manage to contribute to such a reinforcement of the inheritance motive. Governments have the possibility of increasing their own public saving. It is, however, possible that such policies will partly be neutralised through reduced private saving (Barro, 1974).

If there is concern in our generation about what living conditions the generations to come would manage to maintain, the following alternatives are possible:

1. The authorities can, by supporting research and information, make our generation aware of the impact of present-day trends. If these impacts are serious, it may be possible to strengthen the

inheritance motive and increase our collective inheritance to future generations.

2. The government can – mostly in the form of international co-operation – contribute to the preservation of important renewable natural and environmental resources in a productive condition. Such a policy would be most significant in maintaining the ability of the world to produce food and prevent global climatic change. As we said in Chapter 5, this agrees with the type of policy which is recommended by supporters of 'strong sustainability'.

3. The government would be able to direct production of knowledge and awareness so that the technology which is being developed can be characterised to a greater degree as sustainable. This will increase our generation's possibility of compensating future generations for the harm which we inflict on the natural capital we consume. This will also increase the possibility of each generation attaining acceptable living standards without simultaneously causing huge consumption of natural resources and inflicting long-term serious harm on environmental resources.

The conclusion must be, however, that even if the government does have the opportunity of provoking measures which transfer income from our generation to future generations, the chances are not very high. The possibilities are very limited where co-ordinated international action is necessary for such transfer (see Haavelmo and Hansen, 1992).

8.2 Discounting and sustainable resource management

Governments will often have to make decisions about projects which have long-term consequences for access to natural and environmental resources. The implementation of such projects will often have positive effects for the present generation but, as a result of the consumption of natural resources and deterioration of the environment, will have negative consequences for future generations. Also, assume that the effects at any point in time can be considered in a meaningful manner as an increase or reduction in the relevant generation's living conditions. How should such a project be valued by present-day governments? Or, to be slightly more concrete: how should future impacts of the project be discounted so that the effect at different points in time is comparable? The answer to this question is, in fact, dependent on the degree to which the project will have distribution effects for future generations when private savings patterns are taken account, and what opportunities the government has to transfer the income among the generations.

Assume, to begin with, that the savings' patterns of private players neutralise the distribution effects so that compensatory investment takes place without government intervention, and such investment results in a return which is equal to the market interest rate. In this case, the project would benefit all generations if, and only if, the project has positive current values using the market interest rate as the discount rate. In this case, the market rate should be used as the discount rate.

It can, however, be more realistic to assume that the distribution effects of the project would be neutralised only to a small extent through change in the savings' patterns of the private sector. In addition, as we pointed out in Chapter 6, there is some doubt as to whether the compensatory investment which is triggered by changes in private savings patterns would be based on sustainable technology. It would have to be based on such technology if the investment is to be in the interest of future generations; this means that the project itself, even following any changes in the savings' patterns of the private players, would have distribution effects which are unfavourable for future generations. The discount rate used by governments in this case would depend on the opportunity they have for redistributing income between the generations.

If the governments have perfect opportunities for redistribution, the future effects of the project should be discounted using the market rate. As we showed in Chapter 4, a reasonably egalitarian sustainable development in a market system would be characterised by a positive market interest rate which does not actually have to be constant over time. Why should this market rate be used to discount future effects of the project under assessment? With perfect opportunities for redistribution, the government can make alternative investments which yield a return which is equal to the market rate. For this reason, if the project has a positive current value when its effects are discounted by the market rate, compensatory investment would make it possible to realise a situation where all generations prosper. Given the anthroposophic base, the project should in this case be implemented.

The situation is radically different if we make a more realistic assumption and presume that the government lacks perfect opportunities for transfer of income between the generations. In that case, the government would not be able to undertake compensatory investments which yield a return which is equal to the market rate. It would then seem that the best way of transferring income to future genera-tions would be to avoid implementing the project. This would apply even if the project has a positive current value when future impacts are discounted by the market rate.

We have discussed above how democratic governments would wish to decide whether a project with long-term negative environmental effects should be implemented or not. The same sort of reasoning could also be applied for assessment of measures which are intended to make polluters responsible for their pollution, when it has long-term negative consequences for future generations. Governments must balance, when drawing up their measures, the short-term advantages to the polluters of being able to discharge their pollution against the immediate and long-term consequences of the pollution. The discount-ing question must be either implicitly or explicitly part of their deliberations.

We have previously concluded that democratic governments can have increased transfers to future generations as an overall aim, while at the same time, the funds which should be available for the realisa-tion of such a distribution are lacking. This means that when projects which have long-term, negative consequences for the future are con-sidered, the discount rate for these effects is lower than the market rate. To what extent future impacts of such a project should be dis-counted will depend on the strength of the goals about handing down to future generations and the return on the alternative compensatory investment that the government is able to put in place. Because the degree of discounting depends on our democratic governments' will and ability to transfer income between generations, the discount rate used in such project assessment cannot be directly harmonised with the principle of intergenerational justice. Even so, the principles of justice can have indirect significance by influencing the public debate which determines the will of governments to transfer income in the favour of generations to come.

9
The Premises for Ecologically Sustainable Development[*]

9.1 Indicators of sustainable development in Norway

During the first phase of the Project for a Sustainable Economy, an 'Environmental Parliament' was given the responsibility of carving out the premises and demands which would outline nature's ecological framework for human activities (Box 9.1). This framework was to form the basis for the alternative scenarios in the project.

Box 9.1   The 'Environmental Parliament'

The 'Environmental Parliament' consists of members of Friends of the Earth–Norway. The Parliament also has broad ranging contacts with professional biologists and ecologists. In the Parliament's introductory phase, the Project for a Sustainable Economy arranged a consensus conference to select and debate ecological indicators in which ecologists, biologists and economists participated.

The 'Environmental Parliament' arrived at four fundamental principles for management of ecological sustainability:

- preservation of life-supporting systems
- preservation of biodiversity
- sustainable use of renewable resources
- use of non-renewable resources, which does not conflict with the three previous demands.

---

* This chapter builds on Hansen *et al.* (1995).

These basic principles have a lot in common with the pragmatic economic-analytic interpretation of strong sustainability discussed in Chapters 5–8. With this as a basis, the 'Environmental Parliament' then drew up a number of environmental indicators which, in its opinion, encapsulated the links between the Norwegian economy and its ecological impacts in an unambiguous and comprehensive manner.

9.2   Indicator categories

Statistics Norway has divided the environmental indicators into four categories along an axis from influence to impact indicators. The four categories are:

- causal indicators (for example, energy consumption)
- influence indicators (for example, CO_2 emissions)
- impact indicators (for example, changes in radiation bombardment)
- speculative impact indicators (for example, global mean temperatures or ice conditions in the Barents Sea). Figure 9.1 shows the categories along the axis.

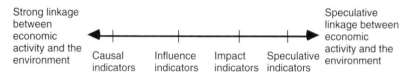

Figure 9.1   Indicator categories

Towards the right along this axis, the linkage between the indicators and economic activity becomes steadily more and more uncertain and speculative. The goal of the 'Environmental Parliament' was to find indicators which measured the condition of the environment directly, while at the same time showing the linkage to human activity as clearly and unambiguously as possible. The indicators should also measure types of environmental change which present a significant risk for the basis of human, and other species', life in the long term. For this reason, the indicators also focus on irreversible and long-term change in the ecosystems.

One problem is the slowness of response. An impact indicator often changes little, even if the activity which is the cause of the problems has ceased or changed character and extent. Environmental toxins often create long-term problems, and levels in the environment can still remain high long after the discharges have been reduced or eliminated. Good impact indicators should, therefore, have relatively short response times if they are to be used to measure the sustainability of the economy and not the extent of former sins. When it is not possible to find impact indicators which satisfy these requirements, the influence indicators and cause indicators give the best measure of the sustainability of the economy. For this reason, the 'Environmental Parliament's' sustainability indicators include a predominance of cause and influence measures.

Technological trends will be significant for the possibility of sustainable development. Energy efficiency programmes will be part of making the production of products and services more efficient in their use of energy and raw material, less polluting and less dependent on physical impact on nature and use of land. The 'Environmental Parliament' chose, therefore, to include a group of technology trend indicators, in addition to those used by Statistics Norway, while at the same time dropping the most speculative indicators.

This gives the following grouping:

• causal indicators
• influence indicators
• impact indicators
• technology development indicators.

After comprehensive consideration, the 'Environmental Parliament' presented 54 indicators which they meant should be included in the basis for assessment of sustainability in an economic development (see Hansen *et al.*, 1995, for a more thorough discussion).

9.3   Statistics Norway: sorting the indicators

The next step on the road towards operationalisation of the sustainability analysis was for Statistics Norway to go through the 'Environmental Parliament's' proposals for premises to see how practically applicable they were for use with the macro-economic models. After having considered them, and having entered into a long dialogue between preservationists and economists, it proved that 26 of the 54 priority indicators from the environmental parliament could be incorporated in one way or another into the project's macro-models (see Table 9.1).

Table 9.1 Indicators which can be transferred to Norwegian macro-economic models

Indicators	Indicator type
Total energy consumption *per capita*	Causal
CO_2 emissions	Influence
Emission of greenhouse gases	Influence
CH_4 emissions	Influence
NO_x emissions	Influence
SO_2 emissions	Influence
NMVOC emissions	Influence
Number of people exposed to air pollution above borderline limits	Impact
Per capita consumption of timber	Causal
Per capita consumption of cement	Causal
Per capita consumption of steel	Causal
Consumption of artificial nutrients	Influence
Per capita consumption of lead	Causal
Consumption of aluminium	Causal
Per capita consumption of copper	Causal
Production of bauxite	Causal
Production of artificial nutrients	Causal
Production of timber	Causal
Number of km of new roads per annum	Influence
Production of cement	Causal
Petrochemical activities in vulnerable areas	Influence
Exploitation of marine resources	Influence
Atomic energy production	Causal
Development of alternative energy sources (bio, wind, wave)	Causal
Gas-fired energy production	Causal
New hydropower projects	Influence

What about the remaining priority indicators?

28 of the indicators found by the 'Environmental Parliament' could not be included in the economic models (see Table 9.2). The Project for a Sustainable Economy has therefore instituted a number of sub-projects which attempt to unify as many as possible of the indicators which fall outside the alternative scenarios. Some of these indicators will be indirectly discussed in Chapters 11–15 and 19–22.

9.4 Ecological demands of the 'Environmental Parliament'

The 'Environmental Parliament' was asked to link concrete demands and goals to the indicators in Table 9.1. The demands for sustainability were fixed, based on natural tolerances, the 'precautionary' principle and fair distribution of 'environmental space'. The 'Environmental

Table 9.2 Indicators which cannot be included in Norwegian macro-economic models

Indicators	Indicator type
Energy investment, raw material consumption and land use needed for the mean consumption of products and services in Norway	Cause
Norway's export of oil and gas measured in terms of amount of CO_2 emitted during combustion	Cause
Specific CO_2 emission per energy unit of oil and gas exported	Cause
Share of gas export which replaces existing coal-fired power stations	Cause
Emission of substances which break down the ozone layer	Influence
Total discharge of 39 priority environmental toxins	Influence
Discharge of N and P to primary recipients	Influence
Annual extent of clear-cut, reforestation, ditching and lumber road-building	Influence
Share of watercourses affected by acid rain which require calcification	Influence
Numbers/density/distribution of reindeer compared with calculated carrying capacity	Influence
Share of agricultural land ploughed in the autumn	Influence
Share of land areas which can be classified as natural, modified, cultivated, built-up and degraded ecosystems	Impact
Share of productive forest areas which are primeval forest, natural forest, cut land, replanted and lumber roads.	Impact
The area of the biotopes which are greatly significant for biodiversity–for example, marine wetlands, river deltas, little affected forest and primeval forest	Impact
Number of threatened and extinct species	Impact
Number of threatened and extinct endemic species	Impact
Number of species with stable and increasing populations, and the number of species in significant decline	Impact
Areas more than 5 km from the nearest road or other technical encroachment (share of 'wilderness' areas)	Impact
Areas more than 500 m from the nearest road or other appreciable physical encroachment	Impact
Share of watercourses/surrounding areas affected by hydropower development	Impact

Table 9.2 *continued*

Indicators	Indicator type
Share of lakes, watercourses and fjord areas which are classified as being polluted (eutrophication and environmental toxins); NIVA's classification is used for eutrophication	Impact
Share of acid watercourses and lakes which have viable fish stocks thanks to calcification	Impact
Levels of environmental toxins in the food chain which, with high level of probability, can be traced back to national sources (eggshell thickness in predatory birds, mercury in fish meat)	Impact
Loss of nutrients and topsoil from agricultural areas	Impact
Funds used for R & D in energy and raw material saving technologies	Technological development
Norway's contribution to innovation within energy and raw material saving technologies	Technological development
Funds used for R & D on alternative energy transmission and production based on renewable resources	Technological development
Norway's contribution to technological innovation on alternative energy transmission and production based on renewable resources	Technological development

Parliament' estimated, on this basis, sustainable consumption and emission levels for Norway.

In practice there are, of course, a number of significant problems concerned with such calculations.

What are the 'limits of nature'?

There is great uncertainty attached to the natural tolerance to many types of pollution and impacts on nature (Box 9.2). There are many excellent methods of calculation for acid rain and components which lead to formation of low-level ozone.

Box 9.2   Components in the sustainability calculations of the 'Environmental Parliament'

1. *Nature's tolerances*: defining the tolerance of nature to encroachment, loss of land, overtaxation of biological resources and ability to assimilate pollution and waste.

2. *'Precautionary' principle*: safety margins for scientific uncertainty are included. As the danger of disastrous, irreversible consequences for life on earth from exceeded tolerances increases, safety margins which also increase must be used at each level of uncertainty as to tolerance/carrying capacity.

3. *Environmental space*: it is possible, based on the link between consumption of raw materials and energy and the pressure on ecosystems, to calculate the sizes of resource removal and polluting emissions which must not be exceeded. This is known as the environmental space.

4. *Fair distribution*: it is decisive for the environmental Parliament that the environmental space is primarily distributed fairly throughout the world. By fixing what is thought to be a fair distribution of the environmental space, it is possible, based on the expected population of Norway and the rest of the world, to calculate Norway's total share of the environmental space at various times in the future.

Consideration for fair distribution of the environmental space is most relevant for resource consumption and emissions which cover domestic consumption. The first stages of the calculation remain the same, while, for example, oil and gas export must also be assessed on the basis of the effects on global and regional consumption levels and emissions.

The same applies to emissions of greenhouse gases, despite the fact that the scientific uncertainty in this area is great. It is often far more difficult to define a tolerance level in scientific terms for discharge of environmental toxins. As far as the various types of land use and encroachment are concerned, the tolerance level is a poorly defined expression. It is difficult to assign demands directly to natural tolerances for aggregated measures – for example, energy consumption. In many cases there is also a possibility of potential environmental harm, and risk analysis is therefore of great relevance (cf. atomic power and the climate debate). In practice it is difficult to avoid the normative aspect when limits are to be determined, and it is more common to speak of 'acceptable harm' and 'acceptable risk' rather than a strictly scientific definition of tolerance.

How strictly should the 'precautionary' principle be interpreted?

This is a normative question, the answer to which depends how great acceptance of risk of various negative environmental consequences is present, and how any short-term benefits can be balanced against expected loss of benefit in the near or remote future. When the ecological demands to sustainability are fixed, they are based on fairly strict interpretation of the 'precautionary' principle. This is especially relevant when there is risk of serious, irreversible change and when the degree of scientific uncertainty is large.

What is fair distribution?

Is fair distribution the same as equal distribution of discharge quotas and resource consumption? Or should the fairness principle be used at a service level (for example, energy consumption vs. energy service level)? More energy is used to maintain a room temperature of 20° C in Norway than in Spain. For simplicity's sake we have based our discussion on equal distribution of the environmental space, as this would seem to be a reasonable interpretation of the fairness requirement of the 'Environmental Parliament'. How reasonable this actually is in relation to such factors as climate, balance of trade, and so on, is worthy of further discussion (for a more detailed discussion of this point, see Hansen *et al.*, 1995).

How quickly should consumption and emissions be reduced?

Equal global distribution of the environmental space sets, in many areas, very stringent requirements to reduced consumption in the rich world (for example, in the use of fossil fuels, minerals and agricultural land). The less time which is to pass before the demand for equal distribution is to be met, and the less time to pass before the demand for limiting resource consumption and discharges which are within the safe limits of the environmental space is met, the more drastic will be the demands for rapid reduction of raw material, energy and land consumption in the rich countries.

The demand for extremely quick reductions in energy consumption and CO_2 emissions can come into conflict with the demand for political realism. Reasons for this include the respect for the balance of trade, national economies and transition and structural adjustment costs (which, as a matter of fact, many people would fervently claim is a question of fairness and therefore belongs with an extended sustainability expression). No one is well served by a rate of change which would destabilise the society economically, politically and socially.

It is, however, important to be clear over which levels of resource consumption and emissions can strictly be called 'sustainable', based on the above assessment and value options. This makes it possible to gain a clear idea of how far away one is from attaining the goals with various choices of framework conditions. We have, therefore, chosen first to fix, and then give the reason for, the demands of the 'Environmental Parliament' on a basis of having attained a sustainable *steady-state* level of emissions and consumption of raw materials and energy in 2030. So, by 2030, with respect to the ecological demands, a world-wide equal distribution of global resources and common goods in accordance with the justice principle of the 'Environmental Parliament' should have been attained.

9.5   The 'Environmental Parliament's' ecological indicators: requirements

Consumption of energy

The ecological demand for reduction in energy consumption set by the 'Environmental Parliament' is based on Goldemberg *et al.* (1988), where the claim is made that energy consumption in the industrialised countries could be reduced by half in the course of a 30–40-year period, while at the same time GNP could increase by around 50 per cent. This could be made possible by implementing energy efficiency measures in their broadest sense and by structural adjustment in consumption patterns towards less energy-intensive products and services. This demand should at least apply to energy efforts behind domestic consumption of products and services. The potential for energy efficiency improvement in power-intensive industry is far less, unless of course, the change is made to secondary raw materials in, for example, aluminium production. In this case, the following dilemma is met: high production levels, on the one hand, increase consumption at a global level; on the other hand, production will pollute more if it takes place in countries which do not base their production on renewable energies such as hydropower, and which perhaps also use more energy-intensive process and technology methods. A milder special requirement for the power-intensive industry, for example, that energy consumption must not increase, could be considered. It is not, however, given that it is better overall for nature to export hydropower in the form of products than as electricity.

The ecological requirement of the 'Environmental Parliament' is:
50 per cent reduction in per capita *energy consumption, compared with 1989 levels, by 2030.*

Consumption and production of mineral raw materials

The environmental impacts of high consumption of raw minerals are linked to extraction, processing, use and waste. The impacts vary highly, with respect to both pollution and encroachment on nature. There is also great variation between the different minerals, some of which are extremely toxic, with long degradation times, where even very modest discharges lead to major and permanent environmental problems. Raw materials which are used in great amounts, including in steel, cement and aluminium production, represent a problem mainly because of the ore-removal involved in mining, and the energy consumption and pollution involved in production.

Norway is a major net exporter of several types of raw minerals, while being a net importer of others. As in energy and CO_2 emissions, it is very important to distinguish between demands to production and to consumption (see the economic sustainability analysis in Section 5.3, p. 54). It is not possible to require the same from extraction and production as it is from the actual domestic consumption of these raw materials.

There is little point in making requirements according to the principle of equal global distribution for production of such energy-intensive raw materials such as aluminium and artificial nutrients. Norway has environmental advantages as a result of its hydropower, and production of energy-intensive raw materials can be a sensible use of Norwegian hydropower schemes, even from an environmental point of view. The same reasoning does not apply to cement, as this production is not based on hydropower. Norway has, however, a lot of raw materials for cement production. Even so, the 'Environmental Parliament' has chosen to set identical demands for consumption and production (calculated in kg *per capita*).

The 'Environmental Parliament's' ecological demand is:
Production of raw aluminium and artificial nutrients must not exceed the present-day level and cement production must be reduced to 80 kg per capita per annum by 2030.

Consumption and production of timber

Timber is regarded in this context as a continental resource, which means that the distribution of consumption quotas is calculated on a European level. The starting point is a level of felling which maintains production over time, combined with felling methods which do not weaken the biological diversity in the forests.

The consumption of timber will be reduced to around 1 cubic metre per head of population by the year 2010, and will then be maintained at the 2010 level: the consumption per head will then change in line with population trends. So liberal a demand (200 per cent of the per-mitted European mean consumption) presumes a consumption of other building materials which is lower than average.

Production of timber is kept at around the same level as today. *The average annual felling level must not exceed 12 million m³.*

Harvesting marine resources

The harvesting level of marine resources is of extreme importance in a sustainability perspective, although it is difficult to fix. For this reason the 'Environmental Parliament' has fixed some overall guidelines for harvesting marine resources. The coastal fishing fleets have a significantly lower energy consumption per fished quantity than the deep-sea fishing fleets. It would therefore seem to be a precondition, seen from a sustainability perspective, that a significant proportion of the fishing quotas be transferred to the coastal fleet.

The 'Environmental Parliament's' ecological demands are:
The harvesting level of marine resources must be conservative with respect to quotas based on the best available management models. The management must aim at avoiding reinforcement of natural fluctuations in fish stocks.

Most of the quotas should be awarded to coastal fishing fleet. It is significantly more energy-efficient and provides more jobs in relation to catch sizes.

Fish farms must be made more secure to avoid fish escaping, and must be located and operated so as to avoid local pollution problems.

Climate gases – particularly CO_2

The UN Climate panel (IPCC) stated in their report (1991):

> We assume with a reasonable degree of certainty that atmospheric concentrations of long life gases only respond slowly to changes in emission levels. By continuing emissions of these gases at present day levels we will expose ourselves to increasing concentrations in the centuries to come. The longer the emissions continue to increase at today's rate, the greater the reductions of concentrations will have to be to stabilise the situation at a given level ... To sta-bilise the concentrations at today's level would require instant reductions of over 60 per cent in emissions of long life gases, while methane will require a reduction of 15 to 20 per cent.

The 'Environmental Parliament's' ecological demands are:
CO_2 emissions must be reduced by 80 per cent of the 1989 level in 2030. Other climate long-life gases must be reduced by 60 per cent of the 1989 level in 2030, while methane must be reduced by 20 per cent of the 1989 level in 2030.

Emissions of NO_x, SO_2 and NMVOC

In addition to the climate gases, the use of fossil fuels results in emis-sion of a number of harmful substances with regional or local impact. It is not possible to focus on all harmful emissions. The model analyses focus on trends in three of them which have been especially central in the pollution question.

Sulphur dioxide (SO_2) has both a regional and local impact. Calculations have shown that close to 95 per cent of the acid rain which falls on Norway originates from sources outside its borders. The main culprit is the combustion of (brown) coal in coal-fired power sta-tions and homes in Great Britain and Germany. Acid rain raises the pH value of fish-stocked lakes, leading to depletion of the stocks, and in areas with little lime in the soil also leads to damage to trees and plants. In certain cases, it can lead to algae flora in freshwater lakes. *The goal is for sulphur tolerances not to be exceeded, even in the most sensitive ecosystems.*

NO_x has very serious effects on health. The Norwegian National Pollution Authority (NPA) estimates that the health of 650 000 Norwegians is at risk from high concentrations of NO_x, mainly from motor traffic. NO_x contributes, in partnership with emissions of liquid organic compounds, to formation of low-level ozone, a harmful substance which plays its part in harming vegetation. So-called 'ozone episodes' during the summer are presumed to contribute to both forest damage and crop loss in agriculture. Emissions today amount to around 220 000 tonnes per annum.

Calculation of an ecosystem's tolerance levels to NO_x is complicated, and the level of uncertainty is great. A reduction in European emissions of an order of 75–90 per cent would be necessary if tolerances are not to be exceeded in the most exposed ecosystems.

Emissions of NMVOC, in partnership with NO_x, leads to formation of low-level ozone during the summer months (photochemical reaction). The tolerances, for both vegetation and health, are being exceeded on a regular basis in southern Norway. The measures needed are technical and not especially expensive.

The 'Environmental Parliament's' ecological demands are:

a   *A reduction in national SO_2 emissions of 90 per cent compared with 1980 by the year 2000 (equivalent Norwegian reductions by 100 per cent 'gap closure').*
    Complete sulphur purification or similar measures for major individual sources in areas where tolerances have been exceeded.
b   *50 per cent reduction of 1990 levels of national NO_x emissions within 2000.*
    90 per cent reduction in NO_x emissions compared with 1990 levels by 2010. Borderline limits for NO_x must not be exceeded in 2000.
    NH_3: 75 per cent reduction of 1990 levels by 2005, maintaining this level until 2030.
c   *75 per cent reduction of NMVOC emissions compared with 1990 levels by 2000.*

Number of people exposed to air pollution which exceeds borderline limits

This is a particularly important indicator. Good air quality is vital for health, and the health aspect is closely connected to quality of life. Good air quality is also of primary importance to the environment.

The 'Environmental Parliament's' ecological demand is:
No-one in Norway should be exposed to air pollution which exceeds the borderline values in 2020. This situation must be maintained in 2030.

Number of km of new roads per annum

Development of the transport infrastructure – especially road building – has a great impact on biodiversity. It reduces vital biotopes, divides up species' living ranges, forms barriers, increases movement through previously untouched areas, and injures game in road accidents (and some of this also applies to rail transport). If increased development of the transport infrastructure in Norway is not to contribute to weakening biodiversity, it is of great importance that new projects do not affect the living areas of endangered species, rare and vulnerable types of nature or other areas which are specially important for biodiversity. Nor should developments affect the large remaining unspoilt natural areas in Norway, or the preservation values of the protected watercourses.

The 'Environmental Parliament's' ecological demands are:
Not even 1 km of new roads should be built before a thorough cost–benefit analysis has been carried out which also includes more environmental effects than those included in present analyses. As of today, noise is the only environmental effect which is calculated in the analyses. The 'Environmental Parliament' will also introduce calculation of land use, emissions to the air and energy consumption. Roads can only be built if such cost/benefit analyses prove that the benefits to society are greater than the costs. The requirement applies whatever the category of road.

Energy production

The following indicators are directly relevant to energy production and will, therefore, result from any demands attached to them. For this reason, no specific demands are attached to these indicators:

- Export of fossil fuels
- CO_2 emissions per energy unit for export of oil and gas
- New hydropower development, km watercourse
- Petrochemical developments in vulnerable areas
- Number of watercourses/surrounding areas affected by hydropower projects.

Hydropower projects are very destructive to the ecosystems of watercourses and lakes. Dried-up rivers, dams and great variations in the water levels of lakes has lead to serious damage both to biodiversity and the productivity of ecosystems.

The 'Environmental Parliament's' ecological demands are:
No change in the number of regulated watercourses from today.

Atomic power

The 'Environmental Parliament's' ecological demand is:
It is a precondition that atomic power is never introduced.

New renewable sources of energy

The production of electrical power by using biomass, solar power, wind and wave energy would depend on energy price trends. These technologies would be able to indicate the back-stop price of energy.

The 'Environmental Parliament's' ecological demands are:

Moderate development of wind and wave power, at a level equivalent to that indicated in the report Energi 2030 [Benestad et. al., 1991] can be permitted, and use of bioenergy can also be increased in line with these scenarios, as long as they do not come into conflict with land use and respect for biodiversity.

Oil and gas production

As of today, oil and gas production in Norway is increasing. Oil production should be limited for price reasons and to restrict the global consumption of fossil fuels. It would also be correct in relation to generations to come. Even if it is a point of discussion how significant the effect is, there can be no doubt that the rapid tempo of Norwegian oil production contributes to delaying energy effeciency improvement in the global economy – by, for example, keeping oil prices low.

The activities of the oil industry also lead to discharges to air and water, and has an inherent risk of major disasters (blow-out, shipwreck, and so on) with potentially catastrophic impact on the environment, especially when oil spills come ashore. The same arguments also apply to a large degree for gas production. The emissions, among them NO_x, are huge, although for the pure gas fields the ecological consequences of a blow-out are less dramatic. Gas is the cleanest of the fossil fuels and if natural gas replaced coal there would be a significant environmental gain in the form of reduced emissions, including SO_2 and $CO_2.$ Natural gas may also have an important part to play on the road leading to technologies based on solar energy. An increase in Norway's gas production can be defended if the gas is used to replace such fuels as coal, oil and uranium, but not if the gas is purely an additional fuel, contributing to an increase in energy consumption in the recipient country (in comparison with an equivalent course which does not include purchase of gas).

Production of gas-fired power in Norway would, however, increase the national emissions of CO_2 and NO_x appreciably, and make it virtually impossible for Norway to meet national goals for these emission components. It must also be empirically probable that Norwegian gas will replace more polluting fuels if increased Norwegian gas production is to be acceptable. The possibility of exploiting waste heat from a power station is less when the probable locations lack potential recipi-ents of the heat and an infrastructure for remote heating systems. Gas-fired power for domestic use will also increase the energy surplus further and make a speedy energy effeciency improvement in Norwegian society more difficult.

The most realistic way of limiting future oil and gas production is to refrain from opening new sectors for oil exploration and production. In general, the risk to the environment from test drilling and opera-tions is greatest when carried out close to the coast and in the vicinity of icy waters. Cold, dark winters and arctic freezing fog in northern waters also increase this risk.

The 'Environmental Parliament's' demands are:

No development of petroleum fields north of the 66th parallel.

No development of fields closer to the coast than at least two days' drift time. The least drift time will vary according to the time of year. The position which is farthest from the coast shall be chosen as the inner limit (drift time to the coast, seabirds' food search area).

No development in the Skagerak (close to the mainland, ecologically overloaded area).

No development of new, pure oil fields with the exception of those already approved.

Development of combined oil/gas fields only if the value of the gas itself makes the development financially viable.

No gas-fired power stations in Norway.

Development of gas fields only if the gas is used to replace existing coal-fired power stations and atomic power stations in other countries.

We have presented the premises of the 'Environmental Parliament' for economic development. There are, however, a range of other factors which must be taken into account in the further consideration of these demands and premises. We will examine them more closely in Chapter 10, where we reformulate the premises to final model demands.

10

From Premises to Parameters[*]

10.1 Introduction

In Chapter 9 we discussed the demands of the 'Environmental Parliament' for a sustainable economy. In this chapter we shall attempt to clarify the economic operationalisation of the demands in the light of the principal arguments of Chapters 3–8. In other words, we apply concrete goals and demands to the indicators which can be included in the macro-economic model calculations.

In those cases where the ecological demands are very uncertain or are speculative, we have tried to translate the demands into possible manageable economic quantities based on the concepts of Safe Minimum Standards (SMS) (see Chapter 6). The main principle is that special natural resources must be preserved and that SMS of renewable natural resources must be the highest instance of control for an otherwise effective resource allocation rule.

It is important to note that the aim of the Project for a Sustainable Economy is to analyse the macro-economic, and some of the environmental, consequences of a heavy restructuring of the Norwegian economy on the basis of the premises of the 'Environmental Parliament'. This again means that, as far as possible, the demands will be maintained even if those to the individual indicators appear to be very strong and it is highly improbable that the model calculations will be able to meet them.

Endogenous and exogenous variables

Endogenous and exogenous variables are distinguished when using models. The values of the endogenous variables are calculated in the

---

[*] This chapter builds on Hansen *et al.* (1995).

model, while the values of the exogenous variables are set by the user. This means that we can have goals for, or make demands on, the values of the endogenous variables, although we have no guarantee that the model will be able to meet them. The values of the endogenous variables are calculated on the basis of the preconditions which are entered into the model, including the values of the exogenous variables being used.

Exogenous variables can be divided into two principal groups. One group is of variables which directly or indirectly can be fixed politically – in other words, variables which can be used as a measure to attain the goals linked with the endogenous variables. The other group contains variables which are determined by other conditions – for example, global economic trends and population trends. This group is stipulated by the model users on the basis of existing data, since it is presumed that the Norwegian government has, for all practical purposes, no influence on them. Twenty-six of the indicators of the 'Environmental Parliament' can be illustrated through the models; 15 of these are endogenous and 11 exogenous.

10.2   Goals which are linked to indicators which can be included as endogenous variables

It is important to note that in connection with models it is possible to link only one goal to endogenous variables. The results from the calculations of these indicators must be correlated with the goals of the 'Environmental Parliament' which are set here.

Of the 15 endogenous variables which the 'Environmental Parliament' discussed in Chapter 9, only the CO_2 goal has been modified. The recommendation of the 'Environmental Parliament' in Chapter 9 is an 80 per cent reduction in CO_2 levels compared with the 1989 level. After having considered this in co-operation with the 'Environmental Parliament', this has been modified to 60 per cent in the model. Even this demand probably exceeds what could be regarded as a realistic political goal in a situation which presupposes 'business as usual' in the rest of the world. A 60 per cent reduction in CO_2 emissions can, however, be defended on the basis of present-day knowledge and the principle of safe minimum standards.

Table 10.1 provides a total overview of the final goals which are linked to the indicators which are included in the model as endogenous variables.

Table 10.1   Goals of indicators which can be included as endogenous variables

Indicator	Goal
Energy consumption *per capita*	50% reduction of 1989 levels by 2030
Steel consumption *per capita*	80% reduction of 1990 levels by 2030
Lead consumption *per capita*	80% reduction of 1990 levels by 2030
Copper consumption *per capita*	80% reduction of 1990 levels by 2030
Artificial nutrient consumption *per capita*	80% reduction of 1990 levels by 2030
Consumption of timber	Reduced to 1 m³ *per capita* by 2010 Total consumption then kept at this level
CO_2 emissions	60% reduction of 1989 levels by 2030
Total emission of greenhouse gases	60% reduction of 1989 levels by 2030
CH_4 emissions	20% reduction of 1989 levels by 2030
SO_2 emissions	Reduced by 90% of 1980 levels by 2000 (equivalent to Norwegian reductions of 100% gap closure)
NO_x emissions	Reduced by 50% of 1990 levels by 2000 and by 90% of 1990 levels by 2010
NH_3 emissions	Reduced by 75% of 1990 levels by 2005, level maintained until 2030
NMVOC emissions	Reduced by 75% of 1990 levels by 2000
Number of people exposed to air pollution	No-one exposed to air pollution over borderline values in 2000, maintained until 2030

10.3   Demands to indicators which can be included in the macro-economic models as exogenous variables

The indicators we indicate here can be directly or indirectly used in the model calculations. In other words, they can be included in the model with the values we fix, and these values join to form the basis for economic trends (and the values of the endogenous variables). It has proved that some of the demands of the 'Environmental Parliament' are difficult to handle separately in the model. This applies to the indicators production of raw aluminium, artificial nutrients and cement, which are all included in large-group sectors. For this reason, we have chosen not to fix the value of these indicators and instead to attempt to interpret trends in the these indicators' sectors in various scenarios.

Nor did we fix production of timber in the models because they would only have a very weak effect on the other results. Instead, we chose to look at how timber production developed as a result of the various measures we examine in the scenarios.

The exploitation of marine resources and the number of kilometres of new roads per annum have not been quantified. However, we maintain the recommendations of the 'Environmental Parliament' as guidelines for the desired direction of development in the fishing industry, and we wish to follow up the Project for a Sustainable Economy with our own additional calculations for this sector, based on the principal calculations we have already carried out. The fishing industry is important, both economically and in terms of the environment and resource management, and has not been examined sufficiently in this project.

If we were to quantify something to enter into the model calculations as number of kilometres of new roads per annum, we would have to attempt to fix a value which would not allow room for investment in new roads, in order to see what the consequences would be. This

Table 10.2   Demands to indicators which can be included as exogenous variables

Indicators	Demands
Production of raw aluminium	Must not exceed present levels
Production of artificial nutrients	Must not exceed present levels
Production of timber	Kept at present levels
	Average annual felling must not exceed 12 million m^3
Production of cement	Must be reduced to 80 kg per person by 2030
Nuclear power	Nuclear power will not be developed in Norway
Alternative energy sources	Moderate development of alternative energy sources (bio, wind, wave power)
Oil and gas fields	No oil and gas field development north of the 66th parallel, no development of fields closer than two days' drift time, no development in the Skagerak
Gas power	No gas-fired power stations will be built in Norway
Hydropower schemes	No new hydropower schemes
Requirements on road-building	No demands have been quantified, but an assessment of the economic consequences of stopping all investment in new roads is desirable
Marine resources	No demands have been quantified, but an assessment of the environmental consequences in the fishing industry sector based on the other demands is desirable

was not done because the results would probably be very difficult to interpret in relation to the goal.

Nor was the indicator 'alternative energy sources' treated in these scenarios. The demand of the 'Environmental Parliament' was 'moderate development of alternative energy sources'. This indicator would have had no effect on the rest of the trends if it had been included.

The remaining demands of the 'Environmental Parliament' on indicators which can be included as exogenous variables are include in the models with the same values as given in Chapter 9 (Table 10.2).

Part III

Questions Calling for More Detailed Sector Analyses

11
Indicators Which Cannot be Handled by the Macro-models*

11.1 Introduction

In Chapter 10 we looked more closely at which demands and goals of the 'Environmental Parliament' we could maintain, either as endogenous or exogenous variables in the macro-economic models. In Chapters 11–15 we will examine the 28 indicators which, for various reasons, cannot be adapted for use by these models. We will also look at a number of subjects which have been focused on in the political debate around sustainable development, but which the 'Environmental Parliament's' indicators do not approach. A central theme here is the environment and employment.

One of the reasons that many of the 'Environmental Parliament's' indicators are not included in the models is that these models were originally designed for the analysis of economic trends for all of Norway as one, where the main emphasis was laid on pure economic quantities. The environmental impacts that the model considers have been only recently included. This, and other structural problems with the models, will be returned to in Chapters 16–18.

Already from its inception, the Project for a Sustainable Economy was quite aware of the limited structure of the models, especially in reacting to local variations throughout Norway. For this reason, the project implemented a number of complementary partial projects in parallel with the alternative scenarios in an attempt to include some of the indicators which would otherwise fall by the wayside.

Many of the limitations in the macro-models are justified by the claim that it is difficult to value environmental effects in an unambiguous,

---

* This chapter builds on Hansen *et al.* (1995).

precise and uniform way. For this reason, the project found it necessary to undertake a partial study of the valuation question. The results of this study were presented in Chapter 7 as a extension of the discussion in Chapter 3.

Based in the four indicator categories we used in Chapter 9, we want to attempt a rough classification of some of the 28 indicators which fall outside the macro-economic models.

11.2   Causal indicators

The causal indicators are well covered by the model runs. The following two indicators remain, and are not included :

- share of gas exports to Europe which replaces existing coal-fired/gas-fired power stations
- energy inputs, consumption of raw materials and land use required to satisfy mean consumption levels of products and services in Norway.

The project has not carried out any partial survey to include these indicators, as there are major quantification problems with both.

As far as oil and gas export is concerned, Norway could have as a principle that all gas exports should replace existing coal-fired/atomic power stations. Unfortunately, as long as we have no opportunity of directing how gas exports are to be used, or even have certain figures to indicate whether such a replacement is taking place, we cannot include such an indicator in the study.

During the spring of 1995 a debate raged in Norway (see, among others, Morten Aserud and Snorre Kverndokk in the financial daily, *Dagens Næringsliv*, 12 April 1995) about whether increased sales of Norwegian gas was an advantage in environmental terms, especially with regard to European CO_2 emissions. Two fundamentally opposing opinions became obvious. One was from the Ministry of Industry and Energy, presented by the minister at the time, Jens Stoltenberg, and the other was promoted by the environmental organisations, partially supported by the director of the EU environmental bureau, Domingo Jiménez-Beltrán (*Aftenposten*, 2 June 1995). Stoltenberg claimed that Norwegian gas was being used instead of coal and nuclear power in Europe, and was therefore contributing to a cleaner environment.

The green organisations, on the other hand, claim that all Norwegian exports come in addition to already existing consumption in Europe. This contributes to pushing prices downward, which leads to increased energy consumption. The reality is probably somewhere between these two extremes. The question is which of these two effects is the stronger. This will, in turn, depend on the price of gas compared with coal and the demands of the energy markets.

If gas is cheaper than coal, consumers would choose between the cheaper alternative, which in an isolated sense would result in reduced emissions. This means that consumers would spend less money when purchasing the same amount of energy. The question then arises as to whether they would exploit this saving by buying more energy or other products.

If gas is more expensive than coal it would not compete with it, but with other, higher-priced sources of energy. There are examples of Norwegian gas outcompeting German alternative sources. There are also examples of Norwegian gas outcompeting Norwegian export of electrical power. If this happens the result would be an increase in CO_2 emissions.

Finally, these effects depend on institutional rigidities which mean that price mechanisms are not allowed to work freely.

If the decision-making authorities in a country institute environmental preferential treatment, it is possible that they would choose to invest in Norwegian gas instead of granting a concession for the construction of a coal-fired power station. On the other hand, special consideration for jobs within the domestic coal industry might mean that consumption of coal was protected. In such a case, gas would more likely be used as a supplement to coal, rather than as a replacement.

The question of whether Norwegian oil and gas exports really contribute to reductions in CO_2 emissions, as claimed by supporters of increased development, would be answered only by a comprehensive empirical study.

There would also be major empirical problems in the measurement of energy investment, raw material consumption and land use behind average consumption of products and services in Norway. We also had to admit that the additional information that we would gain by operationalising this indicator through a partial project was so little that we did not regard a survey as realistic.

11.3   Influence indicators

Influence indicators comprise the group which is mainly relevant for the partial projects. These indicators include 'Discharges of nitrogen and phosphor to primary recipients', which is the indicator which most closely approaches a study. Our partial project, 'Valuing the environment', includes a case study of the North Sea Treaty which is especially concerned with reaching agreement among the countries which border the North Sea about major reductions of polluting discharges. This is also a central feature of our partial study about the environmental impact of a deregulated agriculture and a liberalised trade in agricultural produce. (See also Chapters 13 and 20 for more information.)

Emissions of ozone-depleting substances

Those ozone-depleting substances that we are aware of are covered by the Montreal Protocol, and will be phased out by 1998. There is no relevance for us to devote resources to studying an indicator which will soon be on the way out, and we have chosen not to give priority to this indicator in a partial project.

Discharge of environmental toxins

This is a serious environmental problem, because these toxins have a great accumulative effect. The degradation time for certain of these substances can be as much as 100 years, and we are not yet sure of the extent of the harm they inflict on nature. There is much to indicate that some of these toxins have great influence on human and animal fertility. The Project for a Sustainable Economy has not been able to approach this question in any explicit way.

Other important influence indicators

Such indicators include the number, density and distribution of reindeer, the number of autumn-ploughed fields, annual number of clear-cut forests, ditching and road-building (see Table 9.2). These are important ecological indicators, although they are extremely difficult to treat in a national economic project. The partial study we have about valuation may possibly give us valuable input as to whether it is possible or useful for the process of political decision-making to have the extra knowledge provided in the form of a monetary price for this sort of indicator.

11.4 Impact indicators

Impact indicators are the group of indicators which receives the least even-handed treatment by the models. Most of the indicators have to do with biodiversity and land use. These indicators are especially difficult to value in monetary terms. We attempted to design two of the partial projects to be able to include the impact indicators.

One of these focused on the impact of more liberal international trade in agricultural produce resource effectiveness and biodiversity. This project should encapsulate all of the indicators which have to do with biodiversity and the results are given in Chapters 13 and 20. Another partial project focuses on the extent and distribution of land and resource rent for our most important resources. This partial project should encapsulate the land use part of the impact indicators. The results of this partial project are given in Chapter 23.

11.5   Technological development indicators

Opinions on the opportunities offered by technological development have always been a controversial topic in environmental and development economics. Criticism has been directed at the exogenous technological premise, and claims that endogenous technological growth explains far better the differences in international economic trends and eras (Romer, 1990). Nineteenth-century technological pessimists predicted that nature would be exhausted and we would finally survive at a subsistence level. They overlooked the ability of a society to redevelop human resources and new forms of organisation, in this way making it possible to exploit man-made capital to ease the extraction and processing of natural resources to satisfy human consumption.

The technological optimists base their view on the belief that human beings will always be able to solve the serious problems we meet in our society. Those voices which warn today about pollution, negative accumulative effects and depletion of natural resources are often regarded as technology pessimists.

The 'Environmental Parliament' suggested four technological development indicators which were all linked to technological development in energy or raw material-economising measures.

The project has no partial project which has been directly technology-oriented but the impact of various average percentages of technology improvement was studied in the partial project concerned with whether the various approaches to a domestic Norwegian CO_2 tax would affect domestic public and private transport (see a more detailed discussion of this in Chapter 21).

12
Green Policies and Employment Incentives: Can More Stringent Environmental Demands Lead to 'Double Dividends'?*

12.1   Introduction: taxation and employment

There is a lot to indicate that both environmental and employment questions will be near the top of the political agenda within the near future, both in Norway and in most other countries which share our economic level. It is then tempting to ask whether, and if so, how, the two political areas are linked, perhaps even if they can be influenced simultaneously. This is first of all a question of whether it is possible to follow policies which have putting the environment in focus as their primary goal, while having, at the same time, a positive impact on employment.

The question of whether increased environmental taxation could contribute to general allocation gains through reductions in other, distortionary forms of taxation, has recently assumed a prominent position both in Norwegian public debate and in official Norwegian policy. NOU (*Norwegian Official Report*, 1992:3, 'Towards a More Cost-efficient Environmental Policy') states, for example, that environmental considerations are given a 'price' through increased application of environmental taxation which would also, at the same time, allow room for improved effectiveness of the fiscal system by promoting lower tax rates elsewhere in the economy. In this way, environmental taxation can be seen as an important element in a policy of improving the way the economy works. Even if it is not explicitly stated here, it is implicit that an important element in 'improving the way the economy works', at least in today's situation, is that employment increases. More directly, the former director of the Norwegian Central Bank, Torstein

---

* This chapter builds on Strand (1995a).

Moland, speaking on behalf of the Nordic countries, launched, at the annual conference of the IMF, the concept of combining increased environmental taxation with an equivalent decrease in labour taxation. This takes it for granted that the consequence would be to increase environmental dividends and employment levels at the same time. This view was repeated during his annual speech to the Bank of Norway and Moland's initiative actually goes to the core of the problem which this chapter focuses on.

It could be useful, to begin with, to point out the fairly obvious fact that a measure which has a desired impact on the environment could have either positive or negative effects on employment. The effects on employment of any given measure depend on both the nature and strength of the measure. With regard to the nature of the measure, it would intuitively seem to be easy to understand that the immediate employment effects would be stronger in a case where workers, who were unemployed to begin with, became employed in environmental cleansing projects, than in a case where the same environmental improvement was caused by closing down individual factories.

As far as the strength of the measure is concerned, it is easy enough to understand that requiring a company to reduce polluting discharges, using economically manageable purification measures, will often have more favourable employment effects on the society as a whole than a demand that the company eliminate its discharges completely, which may lead to the company folding. There is, however, not always a clear or unambiguous connection between the immediate impacts of such individual measures and the impact on the economy as a whole, especially if the government is allowed to use other measures to promote its desired policies (for example, using other taxation variables to promote employment).

An important question of principle which will have great significance on the following discussion is also raised. The question is whether the links between environmental change and changes in employment levels are dependent on government policy in other areas – for example, whether the rest of their policies are optimal or not. A single example might be the situation with high levels of unemployment. The government, in principle, could reduce these through other than environmental measures (for example, changes in levels of company taxation, or in the unemployment benefit system, or in the employment services), and these measures would be socio–economically effective. Then imagine that the government (for whatever reason) did not decide to implement such measures, but instead concentrated on implementation of measures that were primarily motivated by environmental concerns, and which had the additional effect of decreasing unemployment figures. (Examples could include clean-up actions and support to industrial environmental investment.) Is it then correct to say that these employment gains are a result of the environmental measures?

The answer to this question actually depends on the answers to other questions, including these two:

1. What sorts of measures are, seen from a total socio–economic viewpoint, most effective with regard to increasing employment: those measures which have an environmental motivation, or those which haven't?
2. Will the government consider implementing the relevant non-environmental measures, or will such measures not be considered?

Only the answer to the first question is interesting in our connection, seen from a traditional welfare theory viewpoint. If the environmentally motivated measures are the most effective, totally speaking, they should be implemented, and the impact on employment attributed to them. If, on the other hand, employment could be increased through other types of measure, which totally speaking are more effective than the environmental measures, then these should be implemented first and the employment gains attributed to them. Based on the traditional theory of economic policy, the only reason for the government not to implement other effective measures would be that they lack information about the impact of the measures. If the government had this information, the measures would have been implemented. (Another problem is, of course, that there could be political or professional disagreement about the impact, or the suitability, of various measures, even when all of the impacts had been studied and published. The argument here presumes that such 'professional controversies' do not occur.)

In practice, question 2 would still be of interest. It quite often happens that measures which professional economists or experts consider to be socio–economically effective are not implemented because they meet political opposition, or because there is a lack of governmental ability or will to implement them (see Buchanan, 1960, 1968; Buchanan and Tulloch, 1962; Niskanen, 1971). In that case an acceptable basis may be to accept as given that such alternative measures would not be implemented, in which case the employment effects of relevant (and potentially realisable) environmental measures can be studied. An important point is that environmental measures can be politically easier to implement than other measures which could also have satisfactory effects on employment.

The following discussion considers the various arguments which have been promoted, and relates the changes in employment levels to changes in environmental policy. First, we will discuss the arguments which cause conflict between improved environment and increased employment levels, followed by the arguments which indicate that both can be promoted at the same time. Then follows the more limited question about whether changes in the taxation system leading to more taxation of environmental benefits, and less on direct taxation of labour, would lead to simultaneous improvement in the environment and increased employment levels.

12.2   The debate about the links between employment and the environment

In the following section we shall attempt to systematise and discuss some of the most important arguments which are presented to support opposing claims:

1.   There is a fundamental conflict between employment and environmental interests; we cannot have more of both but must choose which of these we wish to give priority to.
2.   It is possible to carry out policies, or implement political change which allows both interests to be promoted at the same time.

Potential conflicts between employment and the environment

To begin with, we will systematise and discuss a number of conditions where the case is (at least apparently) that an improved and cleaner environment would necessarily come into conflict with goals of higher levels of employment. This may be called the 'traditional view' – it is not possible to have your cake and eat it too. A choice must be made between environmental benefits on the one hand, and jobs (or perhaps a broader expression such as production or material benefits) on the other. Based on the discussion connected to an optimum economic policy, the traditional view has, in many respects, a high degree of common sense. As long as the economic policy is fundamentally effective, it would not be possible to gain more of all benefits without restructuring policies; if it were possible, it would mean that the original policy was fundamentally flawed. If one now focuses on the possibility of simultaneously improving employment levels and environmental quality, there would be two immediate arguments against those who present this argument as 'proof' that improved environment and more jobs are not simultaneously attainable in practice. The first argument is that there are benefits or variables, other than jobs and the environment, which might suffer at the same time as both job levels and environmental quality improve. In this case there is no reason to believe that government policy was fundamentally flawed. The other argument is that it is far from certain that the policies are basically effective. For the time being, we are discussing cases where the employment and environmental gains do not seem to be able to be attainable simultaneously; we will then extend the discussion to cases where such simultaneous gains are more probable.

1   Development of production or extraction capacity in industries which extract or process natural resources directly

A clear and fairly direct conflict between environmental quality and preservation of resources on the one hand, and employment on the other, is connected to the extraction of non-renewable resources like oil and gas and a number of other minerals, and to development of certain renewable resources like hydropower. The depletion of non-renewable natural resources leads, *per se*, to a reduction of the resource base, and will have a negative effect. The extraction of the resources will also, however carefully the production facilities are designed or the activities are managed, lead to strain on the environment. In oil and gas production it is, in practice, impossible to protect against small, sporadic spills or against the possibility of blow-outs. Another condition is that the increased availability of fossil fuels on world markets leads to greater global consumption of energy (a result of lower prices) and therefore greater local pollution and greenhouse gas emission. In the case of hydropower, the schemes will, in practice, always mean negative encroachments on nature, however the projects are designed. This has become clear through the analysis and decision process behind the 'Collective plan for watercourses' (see Ministry of Environment, 1984, 1987; Carlsen, Strand and Wenstøp, 1993; and others).

The impact on employment of such projects can be divided into two principally different components. The first is that the construction and operation of extraction projects leads to increased employment. These could be called the direct effects. These have been thoroughly studied in a separate study by the Project for a Sustainable Economy (*PBØ Report*, 7, January 1995) and the results are discussed in Chapter 19.

The indirect effects brought by the extraction of the resources come in addition and are the second component as a result of increased income in both the public and private sectors, and from all of the changes in the job market which follow from both the direct effect and the increase in purchasing power. These are far more difficult to esti-mate, although it is clear that they can be appreciable. However, totally speaking, they can go either way. It is, for example, reasonable that the positive effects from increased demand will dominate in a situation where there would otherwise be high levels of unemployment. When there is increased exploitation of resources the main effect may instead be that the direct increase in jobs 'steals' workers from other domestic sectors through an increase in wage levels. The argument also depends on the sort of labour engaged to work on the projects. The more skilled the labour force is, the greater the likelihood of it being employed in other industries, and the less the net effect of the project on employment levels elsewhere in the economy.

2   Other conflicts between land/resource use and industry

The concept here is that preservation of land areas which have been denoted nature reserves, leisure areas, or in some other way protected, can lead to reduced activity within the agriculture, forestry or commer-cial tourism sectors, which would have had a beneficial effect on the surroundings. Seen in a total perspective, it is clear that this has a neg-ative impact on employment. In the case of Norway, this aspect would not seem to be particularly significant, especially because lack of space is no particular problem. It is difficult to imagine that significant further job reductions in Norwegian agriculture would be a direct result of such effects.

3   Reduction in employment levels as a result of higher environmental taxation and/or more stringent quantitative environmental restrictions

If companies have to pay environmental taxation which is not com-pensated for through other forms of taxation relief, it will be quite clear that this increased tax burden will lead to reduced profitability, and the greatest reduction will be in companies with the largest pollu-tion problem. The response of companies to such increases in costs would normally be to reduce production and jobs.

As an alternative to environmental taxation, governments can reduce companies' discharges through quantitative regulation, either aimed at the individual company or in the form of marketable quotas, which either are initially distributed free of charge to companies or are purchased from the state at a market price. Quotas which must be bought at market price will have essentially the same effect on profitability levels and employment as equivalent environmental taxation. Direct quantitative regulation, or marketable quotas which are distributed free of charge according to the individual companies' use of them, will in principle have the same effect as the environmental taxation on companies' production levels and jobs, but will lead to higher profitability as the taxes are not deducted in net terms from the companies. One difference between direct regulation and quotas which are awarded according to their need is that the latter case ensures an effective distribution of the quotas between the companies (continued competition in the quota market), while the former case does not ensure effectiveness (because the government does not often know in advance which companies use their quotas in the most efficient manner). This means more jobs with quantitative regulation than with environmental taxation, as long as the effect of the environmental measures is seen in isolation. The main problem with quantitative regulation in this connection is that regulation, in contrast to taxation, does not provide the state with funds which can lead to reduction in other taxes or charges which might have provided a positive effect on employment. It is, for example, difficult to imagine situations where more restrictive quantitative environmental regulation, taken in isolation, would lead to job creation.

4   Relocation of industry as a result of stricter environmental demands or increased environmental taxation

There is a clear relation to the last point above, as stricter environmental demands can lead to loss of jobs. The mechanism which produces these losses is, however, somewhat different from that described. The main concept is now that competition between countries can force environmental policies which form part of a greater strategy in certain countries to attract industries which would otherwise be located elsewhere. The main point here is not that the total economic activity (in the whole group of countries being examined) is changed, but that the location of the industry within the group is changed (Rauscher, 1994).

This situation is of great potential significance if large numbers of companies are virtually indifferent as to the country in which they will locate their activities. This may well lead to a situation where countries compete for the favours of various companies by creating the best conditions for them in many different ways. One of these competitive factors can be the size of environmental taxes and the degree of restriction on industrial discharges, always presuming that the countries do not agree on a common platform for such taxes and demands. This can create situations with too liberal environmental policies, especially when much of the pollution is cross-border and when the polluting companies are very profitable or create many jobs, making them attractive to host countries.

It should be noted here that this sort of competition between countries when attracting industry is not restricted to environmental variables, but will also apply in a more general sense to the entire company tax burden and the overall conditions the company would face in each of the countries. Environmental policy can be especially attractive as a competitive factor in attracting industry, at the same time as such competition is also especially harmful in this area. This is because one result of international effects of environmental emissions is that the individual countries will have an interest in breaking agreements which limit their own emissions, while other countries gain from the restrictions.

Opportunity for increased employment as a result of environmental measures

The flip side of the coin is the question of whether more environmentally sound policies can lead to higher employment levels. To look at this it is useful to return to the discussion in the first part of this chapter, where we distinguished between situations where the economic policy was effective and situations where it wasn't.

If we assume that economic policies are as effective as possible in areas other than the environment, it is far more difficult to prove that a 'double dividend' could exist than in cases where the economic policy is basically not so effective. We now progress to a discussion of certain situations which lead to a probable 'double dividend' and where there is not necessarily a presupposition of initially effective policy. The first such situation is as follows:

5   *The effect of increased state income through environmental taxation can form a basis for decreases in other, especially distortionary, taxes and especially taxes which may be detrimental to employment levels, so that both the environment and employment situation are improved*

We will just mention this situation, as it will be discussed in more detail below. We shall just note that if there are other taxes which are sufficiently unfortunate for employment, and if the increase in environmental taxation forms a basis for lowering these unfortunate taxes, then a double dividend is possible.

6   More stringent environmental demands can lead to improved exploitation of resources through better employment of waste products, and through the development and manufacture of new products

The concept here is that companies are given an incentive to find out how substances previously regarded as waste products can be exploited commercially; and also how manufacturing technology can be improved so that less waste is discharged and more is reused in production. Such improvements in resource use can create jobs, especially where new production can start on the basis of such innovation. This is discussed in more detail and illustrated later.

7   Technological improvements which save energy and lessen other environmentally challenging processes can lead to savings for industry and total increases in jobs

Reduced use of energy can lead to displacement in economic activity away from the energy sector, which is not particularly-labour intensive towards more labour-intensive industries. This can lead to increases in total employment, both because of the increase in real income which this leads to, and as a result of the redistribution of activity between the industries. An American study by Geller, DeCicco and Laitner (1992) shows, among other things, that increased investment in energy efficiency measures towards 2010 could lead to an increase in total employment in the USA by around 300 000 jobs per year. Norwegian studies, carried out by Statistics Norway (including Bye, Bye and Lorentsen, 1989; Moum, 1992) also indicate that appreciable environmental improvement can be achieved in Norway if activity is transferred from capital-intensive to labour-intensive industries, especially since many capital-intensive industries (including metal and wood-processing industries and the oil sector) are both major energy users and discharge great amounts of other pollutants.

8   Investment in purification technology or other environment-protection technology can itself have positive effects on employment, either directly or through the multiplier effect on other industries

The main point here is that if such investment means that the total level of investment in the economy increases, and there is spare capacity in the economy to start with (involuntary unemployment), then the investment, seen in the light of a traditional macroeconomic analysis will lead to more jobs.

9 Innovation in the development of environmental technology can create separate niches and spin-off effects for industry

First of all, there is a large international market for advanced environmental technology products and services, where Norway can gain market shares through development of skills and technology in this field. Secondly, the technological improvements brought by such research can be used by industry as a whole, perhaps even in situations which have little direct connection with the environment. All in all, this can be a useful situation for employment.

10 Active protection of the environment can create jobs directly

Clean-up actions and recycling of suitable waste can have a benefit on employment, especially for unskilled labour. Many such projects are highly labour-intensive, seldom requiring higher education.

11 The change to more extensive, less environmentally challenging operations in Norwegian agriculture and forestry can lead to more jobs in these industries

In agribusiness some claim that it could be possible to find new and profitable product niches, based on ecological and more resource-friendly operation than today. It is also claimed that a restructuring of the agricultural subsidy system, with fewer capital and product subsidies and more direct labour subsidies, would form the basis for maintained profitability of agricultural labour with far less use of environmental and resource-depleting products. The argument is less powerful in the forestry industry; it is difficult to imagine developments in the direction of more labour-intensive operation without a simultaneous reduction in productivity and therefore profitability.

12 Preservation of natural areas can in itself create a basis for new industry and therefore jobs

Protecting nature reserves and banning such traditional industries as agriculture, forestry, commercial hunting, mineral extraction and hydropower schemes can make areas more attractive for tourism and leisure activities. This can make room for companies which can offer products to relevant groups. If these groups come from outside Norway, the activity will lead to increased export and foreign exchange. This means that we are not looking at pure movement of activity from other sectors. A change from commercial whaling to offering whale safaris is one example of such industrial change.

This is situation 2 turned on its head. It is an empirical question as to which of these two situations is the more important. Note, however, that there are limits to the visit frequency, and therefore the size of the activity, that a 'protected' natural area can provide a basis for and still be called 'protected'.

13   Reduction in harm which results from decreasing levels of pollution can lead to increased productivity and employment

Reduced air pollution can lead to reduced illness in the population which, in its turn, can lead to reduced health costs and increased productivity. This has been proved in epidemiological studies of the link between air pollution and health (see Lave and Seskin, 1977). A more recent Norwegian study is Clench-Aas *et al.* (1989). The reduction can create improvements in welfare which, in certain situations, could be transferred into job increases, especially because the increase in productivity can make it more profitable for companies to increase the workforce. There is, however, somewhat more doubt as to whether any savings in the health sector will lead to higher employment, through a possible reduction in taxation. This doubt is a result of the very labour-intensive nature of the health sector, so that reductions in activity in this sector and matching increases in activity in other sectors would be hardly likely to yield a significant net positive employment effect.

Another effect would be that industrial costs are reduced as a result of lessened damage to structures when air pollution is reduced (see Glomsrød and Rossland, 1988, for a Norwegian study of the link between levels of air pollution and material damage; and also a well known overview study from the USA by Freeman, 1982). Another possible effect is intercompany externalities – in other words, that certain companies use production factors (for example, air or water) which are polluted by other companies. All of these effects can, in principle, lead to employment gains through improvement in company profitability. Few attempts to quantify the effect of such factors have, however, been made in Norway.

Situation 5 will be discussed in more detail later so it needs no further discussion here. Situations 6–13 do, however, require a little more discussion. As far as situations 6, 7, and 9 are concerned, the main argument is that more active environmental policies mean that companies become more easily aware of certain technological or commercial opportunities, which would probably already have been present, but which were not easy to see without the environmental impetus. This is, to a certain extent possible, but it is important to be aware of the limitations of the reasoning. Active environmental policies have not always been necessary to provide employment increases of this type. It is, perhaps, more probable that the argument is valid in situations where the government itself is the prime mover and finances the necessary research and development. There should not be any special reason to expect that industry itself will come up with these ideas, independently of government measures. Ideas from recent growth theory, especially Romer (1990), also indicate that increased public investment in technology development can increase the rate of growth in the economy, and that the growth rate in a deregulated situation would be less than optimum. A factor of this kind would present a case for public subsidy of all types of technological development, including those which are linked to reductions in environmental problems.

As far as situation 8 is concerned, there is an obvious argument against it in that, in a situation with spare capacity, any increase in investment would lead to increases in employment. It must then be proved that environmental investment is either more necessary, in a socio–economic sense, than other investment, or that it leads to greater increases in employment for any given size of investment.

Situations 11 and 12 are, as we have already pointed out, virtually the diametric opposite of the related situation 6 in the list of potential conflicts between the environment and employment. It could appear that situation 11 would have little quantitative significance (especially in the face of the ever-stricter productivity demands on Norwegian agriculture in present-day Europe, within or outside the EU), while the situation will be significant only to the extent that Norway invests in mass tourism based on its unspoilt nature (at least compared with most other European countries). One problem is how 'unspoilt' it will then continue to be.

Situations 10 and 13 are perhaps the most genuine and unapproachable. Measures of the type in situation 10 will have especially favourable effects on employment if they can be mainly implemented in recession periods with high levels of unemployment, and for unskilled labour, which is traditionally hardest hit by such conditions. Situation 13 discusses technological externalities which, if they are reduced, will quite clearly have positive effects on companies' profitability, and therefore possibly positive effects on employment figures. The quantitative effects of such situations are, however, unclear and are an empirical question.

12.3   The possibilities of a 'double dividend' from changes in environmental taxation

Clarification of various 'double dividend' expressions

To begin with, a clarification of precisely what is meant by 'double dividend' will be useful. There are various definitions of the expression, depending on how strictly it is defined (Goulder, 1994a). We shall look a little more closely at two of them which we regard as most useful for the practical political debate.

First of all, presume that a specific tax exists which is built up in such a way that when this tax is reduced, and environmental taxation is increased by the same amount, then such a change in the pattern of taxation leads to efficiency gains for the economy, independent of the valuation of the environmental benefits. This would make it possible to move the tax burden from such a distortive tax over to environmental taxation, in a revenue-neutral way, so that efficiency gains are achieved, even when the possible benefits to the environment are ignored. These latter benefits are a 'bonus', a free benefit or dividend.

As an alternative, we can presume that a reduction in the general level of taxation, taken with equivalent increases in environmental taxation will, in total, lead to efficiency gains independent of the value of the environmental benefit. The difference between this and the first interpretation of the expression is that here it is uninteresting (marginally speaking) which of the other taxes is reduced (there could be a proportional reduction in all taxes), when the environmental taxes are increased: in all cases there will be efficiency gains in the economy. This also means that the government will have a free hand to choose which other tax(es) shall be reduced, allowing them, for example, to reduce the actual tax which will have the greatest benefit on employment.

A significant underlying difference between these two double dividend expressions is that the latter is based on the presumption that the rest of the fiscal system (ignoring environmental tax) is basically optimally designed. This means that all of the taxes have the same distortive effect (marginally speaking) on allocation in the economy. The first interpretation of the expression does not generally presume this, so that the loss of welfare involved in a given tax revenue from taxes, other than environmental taxation, is minimised. There will often be a lack of political will to carry out the desired tax changes, even if these will provide a welfare gain for society as a whole. If it is possible to point out that double dividends of the first type can be achieved through certain changes in the fiscal system, then this opinion, and the communication of it to politicians, would contribute to creating the will to carry out reductions in taxation which are basically especially efficiency-obstructing. If such a double dividend is possible it will, virtually independent of environmental preferences, be difficult to argue against an increase in environmentally related taxes: the increase will, after all, have a positive value for society even when the value of improving the environment was zero. This reasoning presumes, of course, that there is no third group of variables, which people have preferences for, and which may suffer when both the environment and employment improve. Such a possible group of three variables, and their impact, will be discussed below.

Note that the discussion of these two expressions has been based on efficiency improvements in the rest of the economy, and not explicitly on improvements in employment levels as the basis for achievement of a double dividend. Very often, these are both concurrent, in the sense that when a measure or a change in taxation leads to job increases it will, at least in isolation, mean an increase in welfare. This is, however, not always the case – for example, when the workers place a high (marginal) value on leisure, or when other variables (for example, company profits) are reduced when the environment and employment improve simultaneously. In such cases, the double dividends of improved employment and environment conditions, do not necessarily mean a welfare improvement for society as a whole:

The opportunity of achieving a double dividend is through an increase in environmental taxation when the fiscal system and rest of the society's economy are effective.

When we examine the theoretical and practical chances of achieving a double dividend through a change in the fiscal system, we can base it on the second interpretation of the double dividend expression. In this definition, it is presumed that the fiscal system, when the environmental variables and environmental policy are ignored, are effective in a socio–economic sense. It is also presumed that there is no ineffectiveness in the economy as long as the effects of the tax rotation, as well as the possible initial effects of inefficient environmental policy, are ignored. One presumption is, for example, that there is no voluntary unemployment in the economy. This means that from the point of view of socio–economic effectiveness it makes no difference which tax is reduced (marginally speaking) when environmental taxation is increased, because in an 'effective' fiscal system, all of the other taxes have the same distortive effect, marginally speaking, on allocations in the economy.

A number of papers, including Bovenburg and de Mooij (1992, 1994), Bovenburg and van der Ploeg (1992, 1993a, 1993b), and van der Ploeg and Bovenburg (1994) discuss various variations on this theme. One central main result is that when the number of potential jobs increases with wage levels, it is not possible to achieve an increase in employment through change in the fiscal system, leading towards increased taxation on environmental benefits and less tax on labour. The answer to the question as to whether it is possible to achieve a double dividend is, therefore, negative. (For further discussion of the underlying preconditions, see Strand, 1994a.)

A number of empirical studies, mainly from the USA, support the view that when the environment must withstand a greater tax burden, and labour equivalently less, it will lead to increases in the gross costs connected to taxation (by which we mean costs which do not take into account any favourable environmental impact), and therefore to no double dividend. Models which illustrate this for the USA are the DRI and LINK models (see Shackleton *et al.* 1992) and Goulder (1994b, 1994c) and for the EU the GEM-E model (Capros *et al.*, 1994) and the AGE model (Proost and Regemorter, 1998). The only example of possible results which do not support the others are from the Jorgenson and Wilcoxen (1994) model for the USA, where a reduction in gross taxation costs is achieved when increased environmental taxation leads to reduction in taxation of labour or capital. There are few results from other countries; a study by Shah and Larsen (1992) indicates, however, that increases in tax on fossil fuels, which are coupled with equivalent reductions in personal taxation in India, Indonesia, Japan and Pakistan, will always result in increased gross costs as defined above.

Impacts when environmental taxation can be levied on both capital and labour in an effective economy

The reasoning thus far has been founded on the precondition that increased environmental taxation in total would be levied on the production factor labour and not on capital or company profits. In a small, open economy, with free cross-border flow of capital, and with free establishment of industry, this is logically reasonable, as the net capital return in such an economy would be determined in international capital markets and, in balance, would not be affected by fiscal conditions. As we have already pointed out, this is already the result of Bovenburg and van der Ploeg's (1993b) analysis, which we referred to above.

We must, however, emphasise that the precondition that the domestic capital return cannot be affected by domestic fiscal policy in a small, open economy like Norway is a qualified truth. In the first place, the foreign investors' lack of acquaintance with Norwegian conditions and industry, and the equivalent lack of knowledge in potential Norwegian investors about overseas investment opportunities, leads to capital returns in Norway that, at least in the short term, vary fairly independently of overseas returns. In the second place, the return on some Norwegian production capital, which has already been invested and written down, could be high and relatively independent of the return on other capital, which means that this return can deviate from the other return in the long term. This applies especially to the return on capital which is based largely on Norwegian natural resources. In such cases it is reasonable that the return would be dependent on, for example, domestic fiscal policy, both in the short and long term, even in such a small, open economy like Norway's. If, instead, we look at a closed economy, or an economy which places restrictions on business establishment or flow of capital, it would be more typical for the capital return to be affected by changes in the fiscal system (Goulder, 1994a). When those products on which higher tax is levied on an environmental basis (for example, fossil fuels) are used to a great extent in manufacturing industries, and especially in the production of capital goods, then this sort of taxation can lead to a greater drop in the capital return than in rewarding labour.

An important question in this connection is where the marginal allocation loss as a result of an increase in taxes (in this case, the socio–economic loss resulting from an increase in tax of 1 krone) is greatest, either through taxation of labour or of capital. If it is greatest when capital is taxed (something Goulder argues is the case in the USA) then a total allocation gain would be easier to achieve if an increased environmental tax is turned over to labour, at the same time as it makes room for a reduction in capital taxation. Several empirical studies from the USA, including Ballard, Shoven and Whalley (1985) and Jorgenson and Yun (1990), indicate that the marginal allocation loss linked to increased taxation of capital is higher than the equivalent loss linked to taxation of labour. See also Bovenberg and Goulder

(1994) and Goulder (1994b) for further discussion of allocation distortions linked to various forms of taxation in the USA, and Pedersen (1994) for a discussion of the principle.

On the other hand, empirical studies for Norway (for example, Brendemoen and Vennemo, 1993) indicate that taxation which is most directly linked to labour as a production factor (for example, income tax) is high, and can be more distortive than the capital taxation alternative. Capros *et al.* (1994) and Proost and Regemorter (1994) similarly indicate for the EU that the greatest allocation gains are linked to reductions in employers' labour costs. In this case a double dividend would be most likely (under the other preconditions on which Goulder's analysis is based) if an increased environmental taxation could be transferred over to capital and make room for easing taxation levels on labour.

Goulder (1994b) finds, in the case of the American economy, that manufacture of capital products is especially energy-intensive. Since taxes on capital also seem to be especially distortive in the USA, he argues for the view that high energy taxation is not necessarily a good thing – and that it is, at least, probable that this sort of taxation will not yield a double dividend. The opposite may be the case in the Norwegian economy. If the Norwegian situation is that capital taxation is less distortive than most of the remaining taxation, this indicates that increased energy taxation will produce a strong double dividend. This can be the case even though taxation of energy used as a raw material in production is probably fundamentally greater in Norway than in the USA, although perhaps not so very much higher.

If the above reasoning is correct, it would also tend to indicate that it is easier in Norway to achieve a double dividend by increased taxation of energy products used as raw material in production (especially of capital goods) than by increased taxation of energy consumption. Since fuel oils are used to a great extent in production, while petrol is mainly a consumer product, it should therefore be more probable to achieve a double dividend through increase in fuel oil taxation, rather than by a further increase in petroleum tax. The opposite would apply in the USA, in other words increased petroleum taxation seems to be especially attractive there concerning the possibility for a strong double dividend. This is one of the conclusions in the general equilibrium analysis of the USA carried out on behalf of America's Environmental Protection Agency (EPA) (see Roger Brinner *et al.*, 1991). This agrees well with an intuition that petroleum taxes are 'too low' in the USA and possibly 'too high' in Norway, at least compared with other taxes on other energy products. This is also indicated by Brendemoen and Vennemo's (1993) calculations which, in the case of Norway, find that increases in petroleum taxation are less favourable than increases in mineral oil taxation, especially when the estimated values of the reduced externalities are taken into consideration. Note, in addition, that the double dividend expression does not capture the extent of the possible gains from an improved environment which result from less pollution following the heavier taxes. The potential for the achievement of massive environmental gains in this way is quite clearly also greatest in the USA, where petroleum consumption is far higher than in Norway (around three times as high *per capita*).

The optimum level of environmental taxation in an economy with distortive taxation

Thus far, we have said little about how the optimum level of environmental taxation should be determined in an economy that is distortive from the outset as a result of the fiscal system. Assume that the level of the marginal environmental harm linked to the consumption of an energy factor, for example, fossil fuel, is known. The question then is whether, and in that case how, the tax should deviate from Pigou taxes (for a graphic illustration of Pigou taxes, see, Figure 7.1, p. 72) when we have distortive taxation to begin with, and perhaps also have market imperfections. The principle of Pigou taxes was introduced into welfare economies by Pigou (1938), and means that in an economy which otherwise boasts welfare economic efficiency, an optimum tax on an externality (for example, a polluting emission) is equivalent to the marginal harm that the externality causes for other activities in the economy. The expression has another interpretation here, as it presupposes that the economy is, generally speaking, inefficient. The increase in environmental taxation will then also impose losses on the economy, or gains which are not related to the externality itself, and such gains or losses must be included in the cost expression. Two additional effects also occur here in the distortive taxation expression, compared with the situation without distortions.

The first is the income replacement effect of the environmental tax. This means that a higher level of environmental tax will be able to replace other distortive taxes and then in this way lead to the resulting distortion being reduced. This is a tendency when environmental taxation is higher than the Pigou tax. The other is the accumulated tax effect, which consists of the environmental taxation in addition to other distortive taxes, making the remaining taxes more distortive than they otherwise would have been.

For a good discussion of these effects, and of their accumulated impact on the economy, see Parry (1994). The above discussion about the possibility for a double dividend (in the welfare sense) actually concerns whether the income replacement effect can more than balance out the accumulated tax effect, and in that case make the optimum environmental tax exceed the Pigou level. In other words, when you have a strong double dividend through increased environmental taxation, the optimum environmental tax level will exceed the Pigou level, as assessed by the government. This means, at the same time, that when there is no strong double dividend, a correcting environmental tax must have a level lower than the Pigou level. This is easiest to understand if it is first presumed that the environmental tax is set at the Pigou level, so that it is fully corrected for the social environmental harm. When we do not have a strong double dividend, however, the environmental taxation is more distortive than other taxes and should be reduced to a lower level.

The government has a lower assessment of the income and costs of the private sector when we have distortive taxation in an economy. The reason for this is that the state then must use more than NOK 1 of the private sector's income to increase its own income by NOK 1. Assume that state income is increased by NOK X per NOK 1 expense for the private sector, where $0 < X < 1$. If the marginal environmental harm was then D, the state assessment of this would be XD. It would therefore be equivalent to the Pigou level when assessed by the state. In reality, this means that the environmental taxation can well be set at a lower level than the Pigou level, as normally understood (equivalent to D), even in situations where a strong double dividend is possible. This emphasises that the an environmental tax which is lower than the Pigou level (in its normal sense) is the rule, rather than the exception, in a 'second-best' economy with distortive taxation.

Inefficiency in the economy: the effects of imperfections in the labour market

So far, all the discussion has centred on cases where adaptation to the economy is presumed to be optimum (when environmental externalities are ignored), to the extent that there are no inefficiencies other than those brought about by the fiscal system itself. The fiscal system can create inefficiency when all taxes (apart from environmental taxes) are equally distortive (something which was implicit in the second double dividend expression), or where some taxes are fundamentally more distortive than others (the first double dividend expression).

This presupposes that the government, through implementation of certain measures, is able and willing to counteract other forms of inefficiency in the economy. This requires both that the government has a sufficient number of effective measures available and that they are, in fact, used.

In practice, both of these preconditions can often be questioned, for three reasons:

1. The number of independent measures available to the government is often fewer than the number of goals, making efficient goal attainment in the case of all endogenous variables impossible.
2. An effective abolition of allocation distortions which result from market imperfections could, in itself, demand a use of measures which exceed the possible areas for use of the measures which are actually available; for example, it could be demanded that limited taxation revenues are used to subsidise certain activities.
3. The measures that the government actually has at its disposal are often not fully exploited, both for practical and political reasons.

To put it another way: In practice, governments have never managed to attain a 'conditional or second-best optimum' even when environmental policy is ignored. In Norway (and even more in other Western countries) today, for example, there are unemployment levels which most people agree cannot be part of a conditional optimum. It could appear to be tempting to base a view on the practical situation when discussing (conditional) optimum environmental policy and the possibility of a double dividend, and not on a purely hypothetical situation in which, all in all, policy is presumed to be conditionally optimum.

Analytical problems arise when one moves away from relatively safe analytical ground: that is, optimum fiscal theory, where conditional optimality is presumed. Very much will depend upon exactly in which way the policy is not fundamentally optimum and here, at least in theory, there are a virtually unlimited number of opportunities. Another problem is that very little existing literature discusses such cases analytically. Samples of the problems which we have discussed, and the results in the various cases, are now examined.

Bovenberg and van der Ploeg (1993c) discuss one case which presumes that workers' real wages are fixed, and that there is no involuntary unemployment. They presume companies which operate in a situation of free competition with three production factors – that is, labour, nature (which results in pollution) and a fixed factor with a constant supply. They first examine the situation where the fiscal system is not fundamentally optimum. It appears that, when environmental taxation at the outset is sufficiently low, a triple dividend is possible in which not just employment and the environment, but also company profits, are improved when the tax system leans towards greater environmental taxation and less direct labour taxation. If the tax system is fundamentally optimum (a situation which is also studied by Bovenberg and van der Ploeg, 1993a), then such triple dividends are excluded. The possibility of a double dividend is, however, still possible, where both employment and the environment can be improved by a rearrangement of the fiscal system to one with relatively less labour taxation. The increase in employment will then be attained at the expense of a reduction in return of the fixed factor (in other words, the profit). The greater the proportion of added value which can be attributed to the fixed factor, the better the possibility, and there is a high degree of substitution between natural resources and labour. It is possible to say that the fixed factor 'pays for' the increase in employment.

Another possible approach to the question of a possible double dividend is to base the discussion more directly on concrete types of inefficiency in the employment market, where the reasons for any involuntary unemployment are specified in more detail, and there is not just a general presumption of fixed actual wages, as in Bovenberg and van der Ploeg (1993a, 1993c). This allows several angles of approach. One is to presume that wages are fixed by trade unions, or by unions and employers in partnership. Another is to assume that wages are unilaterally fixed by employers, but that there are so-called 'effectiveness wages', where wage levels in balance are higher than the market level, because productivity increases in line with wage levels. A third possibility is to presume bilateral wage negotiations, between individual workers and employers, without necessarily involving the unions, and where unemployed workers and employers search for each other in the market place.

The literature which relates such labour market questions directly to the question of a double dividend is, however, very sparse. Some works (Strand 1992a, 1992b, 1993b, 1994a) can, however, provide a primary approach to some such problems. All of these theoretical works presume that industrial environmental discharges are affected by the efforts of the employees to prevent them.

It must, however, be emphasised that research in these areas is in its infancy, and that very much still remains before there is fairly certain theoretical knowledge of the impacts of specific imperfections in the labour market on the possibility of a double dividend. In addition, the role played by similar imperfections in other markets, including the capital, credit and research markets also needs to be clarified.

13
The Environmental Impact of Deregulated Agriculture[*]

13.1 Introduction: structural changes and deregulation

Agriculture is a regulated economic sector. Financial support for the agricultural industry consists of protection against foreign competition (import duties) and domestic subsidies. This chapter discusses the consequences of a reduction in import duties and/or subsidies for the environment and resource use. Such deregulation means that Norwegian farmers must compete to an increasing degree with foreign producers, leading to an internationalisation of Norwegian agriculture.

The agricultural sector has been deregulated in many countries over the past few years. The Uruguay round of the GATT (General Agreement on Tariffs and Trade) negotiations was also, to a very great extent, concerned with reduced agricultural subsidies and increased international market access for agricultural produce. The Norwegian Parliament, through its ratification of the GATT agreement and Parliamentary Bill No. 8 (1992–3), Agriculture in the Process of Development, has given its approval to deregulation of Norwegian agriculture. Central elements include lower cost levels in agriculture, greater emphasis on market opportunities, more competitive distribution of agricultural produce and reducing the gap between Norwegian and international prices for agricultural produce.

Post-war Norwegian agriculture has been put through huge structural changes. In 1950, there were around 250 000 agricultural units (farms, smallholdings, and so on) in Norway, while there are only around 90 000 left today. Labour input has been reduced even more than these figures would indicate. Whereas most farms in 1950 had a diversified

---

* This chapter builds mainly on Rickertsen, Holand and Rystad (1995).

spectrum of outputs, most today have specialised in only one or very few crops or outputs.

These structural changes have taken place despite the very high level of agricultural subsidies and virtually full protection against foreign competition. Two significant reasons for the changes are technological developments and high opportunity cost of labour.

Regional specialisation has also taken place, with grain being grown in the lowland districts of eastern Norway and the Trøndelag region, and dairy and sheep farms in the rest of the country. This regional specialisation is partly a result of the so-called 'canalisation policy' which was implemented in 1950–1 (Vatn, 1984). This ensured a high price for grain throughout the country coupled with especially high subsidies in animal farming outside the best grain areas.

Deregulation of agriculture will lead to new changes in agricultural policy and new transformation of agriculture.

The Project for a Sustainable Economy focuses on the following questions:

(a)   What impact does deregulation have on labour use per unit produced?
(b)   What impact does deregulation have on the use of non-renewable and polluting resources per produced unit?
(c)   What impact does deregulation have on agricultural pollution?
(d)   What consequences will deregulation have for biodiversity in the agricultural countryside and in surrounding areas?

For reasons of limited capacity and computational ease we will look only at resource use in milk and grain production. Milk and grain are, by far, most important in Norwegian agriculture, with regard to income, land use, agricultural pollution, the agricultural countryside, and biodiversity (Budget Committee, 1993, p. 48).

The analyses are based in the causal linkages which are indicated in Figure 13.1. Technological developments in agriculture are have been enormous in the post-war period and have contributed to significant increases in labour and land productivity. The demand for agricultural produce has also increased in line with increases in population. Any future increase in the purchasing power of the most impoverished parts of the world population will also increase future demand. Agricultural land is also being exposed to great change though soil erosion, changes in water table levels, and so on. Other factors, including general economic trends, would be significant for agricultural

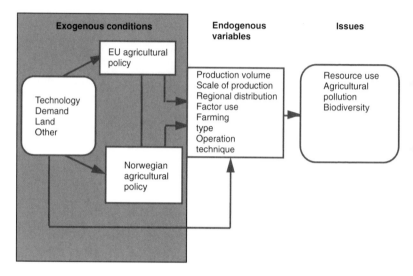

Figure 13.1   Analysis model for Norwegian agriculture

policy, and would be decisive for the development of (opportunity cost) of labour and other production factors in agriculture.

The endogenous variables are production volume, scale of production, regional distribution, factor use, farming type and operational techniques. Any changes in one or more of these factors which results from deregulation of agriculture may lead to changes in resource use, agricultural pollution or biodiversity. By 'production volume' we mean the total production of various agricultural produce, and 'scale of production' means the size of the agricultural units from which the produce comes. The 'regional distribution' tells us where the produce comes from, while the 'factor use' tells us how much of the various production factors is used. 'Farming type' is linked to the degree of specialisation in the production – in other words, whether the individual agricultural unit produces one or more products. The choice of 'operational technique' – that is, use of machines and other equipment – also influences the land use structure of the individual farm.

The different economic models for the agricultural sector can tell us a lot about any impact of the varying degrees of deregulation on certain of the endogenous variables in Figure 13.1. The model results which are presented in the Project for a Sustainable Economy can tell us much about the consequences for production volumes and production scale from policy changes. The model calculations also tell us something about regional distribution and factor use. The models tell us little about farm types and farming techniques; however, farming techniques and type are extremely important factors in an analysis of any consequences for agricultural pollution and biodiversity. For this reason, any consequences of deregulation on choice of technique and farm type (degree of specialisation) are assessed on a more speculative basis. This will indicate only possible consequences of deregulation on pollution and biodiversity.

In an ideal situation, these effects should have been analysed within the framework of a national macro-model (for example, MSG or MODAG). Agriculture is, however, so heavily simplified (aggregated) in these national planning tools that it is not possible to answer the questions that the 'Environmental Parliament' is concerned with in this study, and which this chapter focuses on.

For this reason, two sector specific models are used to analyse the impact on Norwegian agriculture of various degrees of deregulation. JORDMOD is a quadratic programming model which has been developed at the Foundation for Social and Industrial Research (SNF) and the Norwegian Institute for Agricultural Economic Research (NILF) (Box 13.1). NAP (Norwegian Agricultural Policy) is a dynamic econometric model which has been developed at the Institute for Economics and Social Studies (IØS) at the Norwegian College of Agriculture and SNF. The results of these two models are described and discussed in Chapter 20.

Results from these sorts of model are inherently uncertain. This is because such models represent much simplified pictures of reality. The theoretical characteristics and limitations of the models have therefore been discussed in detail in the Project for a Sustainable Economy, *PBØ Report* 11, (August 1995) on which this chapter is based.

Box 13.1   The partial equilibrium model JORDMOD

JORDMOD is the most comprehensive economic model for analysis of the Norwegian agricultural sector. The model was first developed at SNF (Foundation for Social and Industrial Research) and has been further refined at NILF (Norwegian Institute for Agricultural Economic Research). Both institutions have used the model to analyse the effects of various forms of change of domestic agricultural subsidy and liberalisation of agricultural trade.

Box 13.1 *continued*

JORDMOD is a partial equilibrium model which says something about production, demand, net import, employment, land use, capital size, agricultural subsidies, socio–economic profit and industry structures. Deregulation of agriculture also influences other important conditions, especially environmental and pollution problems. JORDMOD, and other sector models, say little about these problems.

A model like this is basically best suited to a small sector in a small country, for example the Norwegian agricultural sector. This is because a small sector influences other sectors to only a very limited degree. Major change in a small sector will, however, influence other sectors. A large and relatively rapid loss of jobs in the agricultural sector is one factor which will affect other Norwegian sectors. Partial models ignore this, and also take prices of other products and income levels as given. In the case of changes which affect several sectors, we get intersectoral cross and feedback effects which affect prices. For a more detailed description, see the Project for a Sustainable Economy, *Report* 10.

13.2   Future agricultural policy

Developments in Norwegian agricultural policy depend to a very great extent on the political will to support Norwegian agriculture. The Norwegian Parliament, as we have already pointed out, has decided that Norwegian agriculture will be deregulated. It is, however, an open question as to how comprehensive this deregulation will be. Three options are regarded as inapplicable in the short term. The first is that Norway said 'no' to membership of the European Union (EU) in its referendum on 28 November 1994. For this reason, the Common Agricultural Policy (CAP) is not on the agenda, nor is its impact on Norwegian agriculture in the event of Norway joining the EU. Secondly, an assessment of a complete deregulation of Norwegian agriculture is not regarded as viable. This option has little or no support in Parliament. Thirdly, agricultural policy will not continue on the same lines as before. One of the reasons for this is the various international agreements, including GATT and to a certain extent the EEA (European Economic Agreement) which restrict the actions of the

Norwegian government. Parliamentary Bill 8 (1992–3) also signals that the government wishes a certain deregulation of the agricultural sector.

The European Economic Agreement

The EEA was signed in May 1992. The aim of the agreement was the establishment of a homogeneous European partnership which included the European Free Trade Association (EFTA) and EU countries. The agreement secures the rights of the EFTA countries to participate in the EU single market. There is dispensation from the EEA for most Norwegian agricultural produce. Norway must allow duty-free admission for 61 Mediterranean products (grapes, peaches, citrus fruits, and so on) from the EU and remove import restrictions on most types of cut flowers. In return, Norway receives duty-free, although relatively small, export quotas for cauliflower, broccoli, Chinese leaf, cherries, plums and strawberries. In addition, there is a comprehensive programme of harmonisation of provisions and technical regulations in the agricultural sector (Dånmark, 1992). This harmonisation leads to increased competition in the market for products and may lead to reduced prices in, for example, agricultural machinery.

GATT

The Uruguay round of the GATT negotiations started in 1986, and was concluded in December 1993. The Uruguay round resulted in the inclusion of agriculture in the GATT regimen and acceptance for monitoring the agreement by the newly established World Trade Organisation (WTO). The agreement means, among other things, that international market access for food products must increase and that the level of subsidies to agriculture must be reduced. This especially applies to export subsidies. These subsidies have especially distorting effects on world trade, and so the agreement contains special provisions for reductions. Both the extent of export subsidies and the amount of produce covered by subsidies must be reduced. The greatest reductions, in percentage terms, must take place in the industrialised countries, where export subsidies must be reduced by 36 per cent and the amount by 21 per cent during the period of the agreement (six years from the start-up date). A number of forms of subsidy are, however, excluded. This applies to direct government funding, which has relatively little effect on the amount produced and international trade.

Norway, in common with most industrialised countries, has already reduced the level of subsidies in the agricultural sector to a considerable extent compared with the reference period(s) and it will not, therefore, be necessary to reduce the level of subsidies in the near future as much as the minimum requirements.

Parliamentary Bill 8 (1992–3): Agriculture in the Process of Development

The bill provided signals about the direction of agricultural policy in the years to come. The market will lead the adaptation of Norwegian agriculture to a far greater extent than it does today. This means that the costs must be reduced, allowing Norwegian agriculture to compete better in a deregulated domestic market. Among the measures to improve competitiveness will be the necessity for governments to 'accept' larger farm units, especially in grain cultivation in the central lowland districts. The distribution network must be made more efficient. Governments must also look more closely at market opportunities when they draw up future agricultural policy.

According to the bill, the government must emphasise the preservation of nature and protect the environment. The bill signals that the measures must be made more targeted, so that production of common goods is ensured and a contribution is made to production taking place in an environmentally sound manner. It also points out that labour-intensive production in areas where there are few alternative sources of employment must be given priority. Great emphasis is still placed on retaining most of the present agricultural land in use.

Illegal use of growth hormones is a problem in many countries. Opening Norwegian borders could lead to pressure on Norway to use these hormones. Liberalised import will also increase the chances of import of hormone-treated meat and milk. Animal health is, in general, good in Norway, a result of well planned health schemes, small animal stocks, the Norwegian climate, and the stringent animal import restrictions. An opening for increased import of, for example, live animals could introduce diseases not at present found in Norwegian livestock. Increased herd sizes might also increase the spread of disease and salmonella bacteria, increasing the costs of control and use of antibiotics. A final concern related to liberalised agriculture trade is that of unpredictable environmental impacts from alien species.

Many people will also be concerned that increased international trade will lead to more transport. This does not apply to milk. The transport needs of Norwegian dairy production are great: farms are spread widely and the milk production takes place a long way from the consumer. There is, for example, very little dairy farming in eastern Norway, which means that the milk must be transported in from other parts of the country. The distance from the heavily populated Oslo area to the milk-producing districts around Trondheim and Stavanger is as great as the distance to southern Sweden and Jutland in Denmark.

As far as agricultural policy measures are concerned, general price subsidies are to be reduced, especially the price of grain. Openings are also created for increased use of the price mechanism to reduce milk production levels. The reduced prices will partially be compensated for by increased use of subsidies on production factors which are especially desirable. This applies especially to land and labour.

13.3   Resource use

The Project for a Sustainable Economy has investigated the actual resource use per produced unit in various Norwegian districts and on farms of varying sizes. We are especially interested in the use of resources which are presumed to lead to pollution and other environmental problems. For this reason, we have mainly emphasised the use of artificial nutrients, pesticides, fuel, machines and medication in dairy and grain production. Our analyses are based on farm practice research data which has been gathered by the NILF. These consist of annual accounting data from more than 1000 farms. The farms included in the project are, in principle, representative (for farms where a significant proportion of income comes from the farm), taking into account the distribution into regions, farming types and size groups.

Some of the farms in the research material do not produce just one agricultural product. If there are systematic differences between the groups we compare with regard to the relative extent of such side-products (where there also are different sorts of production factors), a major part of any proved difference in factor use will be a result of different sizes of other agricultural products. This is one of the errors we attempt to avoid by using control variables.

13.4   Agricultural pollution

A more detailed analysis of the impacts of deregulation on nitrogen (N) and phosphorus (P) compounds pollution from Norwegian agriculture requires specialist knowledge about the changes in agriculture which by far exceeds the degree of detail in the MODAG and MSG macro-models. For this reason, the Project for a Sustainable Economy has carried out a separate partial sector study of this question. The main emphasis is placed on run-off of N-salts, but the emission of various N containing gases and topsoil erosion has also been studied where we have reason to believe that deregulation may have an impact on the extent of such pollution.

In small amounts, run-off of nitrogen and phosphorus nutrients can have a beneficial effect on the environment (fertiliser effect). When large amounts of nutrient salts are discharged they become a problem. This problem may be:

- Local – that is, pollution of watercourses/sources which run through the agricultural district.
- Regional – for example, increased discharge of nutrients to the North Sea basin.

Emission of nitrogenous gases from agriculture can also have various effects:

- Locally, when some of the nitrogenous gases are dissolved in rain water and precipitate locally. This will mean increased application of nitrogenous salts and will also pollute the environment.
- Regionally, because some of the nitrogenous gases are transported through the air for quite a long way before they are dissolved by rainwater.
- Globally, by certain nitrogenous gases (NO_x and nitrous oxide), the so-called 'climate gases' which participate in the greenhouse effect.

The Norwegian government has been aware of all of these problems, but there has been particular attention paid to the discharge of nutrients into the North Sea, because, among other things Norway has signed the North Sea Agreement. This agreement obliges us 'to implement effective national measures directed at reducing discharge of nutrients (by about 50 per cent between 1985 and 1995) into areas where it is probable that such discharges, directly or indirectly, would lead to pollution (of the North Sea)' (North Sea Agreement, 1987).

We shall take a closer look at how the following types of change in production could affect the level of pollution:

(a) Reduced intensity in plant cultivation.
(b) Reduced area of marketable products in plant cultivation.
(c) Reduced livestock production.
(d) Substitution effect: lower concentrated feed prices will, relatively speaking, lead to increased use of concentrates and reduced use of rough feed.
(e) Structural change: deregulation of agriculture will contribute to structural rationalisation in agriculture taking place more quickly than it otherwise would have done.
(f) Regional specialisation of production: changes in the exogenous conditions could, in the long run, lead to a significant change in the regional distribution of agricultural production.
(g) Degree of specialisation on the individual farm.

The results of this analysis are presented in Chapter 20.

13.5 Environmental problems: competition in dairy and livestock production[1]

The most important source of pollution in the dairy industry is run-off of nitrogen salts from livestock manure. In the EU, livestock production is very intensive, for example, in the Netherlands, and several regions will be required to reduce nitrogen discharges through implementation of EU's nitrate directive (Leuck, 1994). In several agricultural districts the levels of nitrate in drinking water exceed that fixed in the EU's 1980 Drinking Water Directive. The equivalent problems in Norway are relatively minor (Forsell, 1992).

Attempts at regulation of agricultural pollution will affect production levels and localisation of production, price levels and farmers' income. The differences between private and socio–economic comparative merits means that trading patterns will change if producers are also influenced to take into account the negative external effects of their production. At present, there is a clear difference between cost levels in Norwegian dairy production and the nearest EU countries (for example, Denmark and the Netherlands). It is reasonable to believe that this difference in cost will lessen (*ceteris paribus*) if the EU introduces more stringent environmental regulation to combat agricultural pollution. Norway will also emerge better from the competition with its neighbours than it would in a deregulated market where environmental costs are not included in the calculation. Certain EU regions have far higher levels of agricultural pollution than Norway; it will not therefore be necessary for Norway to impose as stringent environmental restrictions on its producers in order to regulate agricultural pollution.

In Chapter 20 we compare dairy and livestock production in Norway and the EU by means of various economic analysis models and statistics. An econometric model for milk production in Norway tells us something about the price sensitivity of the producers and their reactions to quota systems and other political measures. The model can also tell us something about technological change over time. Combined with models for EU agriculture, comparisons of cost data and surveys of the nitrogen discharge problem in Norway and the EU, we analyse how environmental regulation will affect the relative competitive situation.

13.6   Taking biodiversity into account

Norway's obligations under the Biodiversity Convention

The Biodiversity Convention (UNEP, 1992) was adopted in the summer of 1992. Norway signed and ratified this convention. This means that it is obliged to preserve biological diversity, as well as undertake sustainable exploitation and use of biological resources.

The convention compels Norway to develop national strategies and measures for preservation and sustainable use, as well as integrating relevant measures in sectorial and inter sectoral plans. Individual sectors, mainly divided up along ministerial lines, have been given responsibility for following up the intention of the convention and developing national action plans, with the Ministry of Environment as the co-ordinator.

For the authorities, especially the agricultural authorities, it is necessary to develop a tool for the effective assessment of the impacts of deregulation compared with the goal of preserving biodiversity in agricultural districts. This question was virtually untouched in the Ministry of Agriculture's draft action plan for the preservation and sustainable use of biological diversity, sent out for comment (Ministry of Agriculture, 1994). This is clearly a sustainability dimension which the macro-economic analysis tool used in the work on the Long-term Programme (LTP) could not analyse. This is also the same tool used in Norwegian macro-economic scenarios. If we are to be able to say anything about the effects of deregulation on biodiversity we will have to use specialised analyses of the role of the agricultural sector as a supplement to our macro-models.

A brief definition of 'biodiversity' is variation in living organisms. In the convention it is defined as: '*Variability in living organisms, including terrestrial, marine and other ecosystems, as well as the ecological complexes that they are part of; including diversity within species, at the species level and the ecosystem level.*'

The expression 'biological diversity' includes all forms of life, and is therefore difficult to apply in management and research applications. How biodiversity should be measured and the significance of biodiversity for the function and production of ecosystems are both subjects of debate.

In Norway, as in other countries, there are large gaps in the documentation of biodiversity. This is especially so at the genus level, but also at the species level. Our knowledge of the structure and function of terrestrial ecosystems is also sadly lacking. If we admit this, then limiting ourselves to the species and society level is necessary for further discussion.

Norway has a limited diversity of species, seen against global levels, and there are few species which only occur in Norway. The relatively low diversity of species is partly because the Scandinavian peninsula was covered by ice until only 10 000 years ago, partly because parts of Norway are difficult to access, and partly because of Norway's geographical location in the far north.

The list of threatened species in Norway covers more than 2000 species, most of them within the fungi and insect groups (DN, 1992c). Many of the species which are classified as vulnerable and rare are so because of their presence in Norway. They are often at the edge of their dissemination zones; northern parts of Norway are on the southern edge of the arctic zone, southern parts of Norway are on the northern edge of the nemoral zone and Norway is on the western edge of the boreal belt.

Agriculture as a threat to, and a condition for, biodiversity

No overview has been prepared for Norway which shows the quantitative distribution of factors which threaten the flora and fauna. It is, however, accepted that around half of the total number of threatened species are forest lifeforms (DN, 1992c). In Sweden and Finland agriculture, after forestry, is regarded as the second most important threat factor to biodiversity (Ministry of Environment, Sweden, 1992). In Denmark, where cultivated land occupies 70 per cent of the country, farming is regarded as the greatest threat (Bramsnæs, 1991). In Norway, it is thought that few species have disappeared as a result of human activity (DN, 1992c). The threat to biodiversity is primarily directed at the destruction and division of habitats, changes in operation in farming and forestry and pollution (DN 1992c).

The consequences of different forms of agriculture on diversity is poorly documented, on both a national and an international level. This is largely owing to a lack of systematic time series of species stocks in agricultural districts.

Agricultural land furnishes the basis for many habitats which deviate from nature, and which provide many culture-dependent species with excellent living conditions. We are fairly well aware of which plant species are connected to these habitats, as well as which birds and mammals can use these areas in search for food and nesting. It has been shown that most of the terrestrial insect fauna, which make up nearly 50 per cent of all of the species in Norway, are connected to agricultural land (Hågvar 1991). This shows that the agricultural land, which makes up around 3 per cent of Norway's land mass, is a key area, a 'hot spot' (Prendergast *et al.*, 1993) for biological diversity. There are two development tendencies which are thought to negatively affect the function of agricultural land as 'oases' in the landscape:

1. Marginal agricultural land taken out of operation.
2. Intensification of farming in the best agricultural areas.

Marginalisation and intensivation have taken place and continue in parallel, leading to cessation of use as agricultural land in areas where, seen in the light of biodiversity, it would have been more desirable to have continued cultivation. At the same time, large agricultural areas have 'shaved away' the diversity of the landscape element. This has, in its turn, had negative effects on the living conditions of both flora and fauna.

Svensson and Wigren (1986) claimed that floral diversity increased until the beginning of this century in Sweden's agricultural districts, and then showed a marked decline. There is reason to believe that there has been an equivalent, although less dramatic, development in Norway. The number of species of plants on agricultural land has, however, increased throughout the post-war period (Ministry of Environment, 1994); this increase is connected to species which are spread by human activity.

Intensive agriculture is not unconditionally negative for diversity. Long periods of cultural influence have created new living conditions for many species. Disturbance – in this case, in the form of farming – is often a precondition for species which are adapted to unstable environments and early succession phases able to compete with other species with other survival strategies. The degree, frequency and continuity of the disturbance is, however, important for the survival and adaptation of the plants (Steen, 1980). It is presumed that a milder degree of disturbance gives greatest species diversity (Steen, 1980); this is in line with the general 'competition–stress–disturbance' model developed by Grime (1979).

The number of threatened/vulnerable/rare species of plants and animals in agricultural countryside was estimated at around 300 in the Ministry of Environment draft for the partial plan for biological diversity (1994), while around 600 species are regarded as in need of attention.

It is thought that especially changed forms of operation, use of fertiliser and cessation of cultivation of traditional farming countryside has had a negative impact on the diversity of species in agricultural countryside (DN, 1992c). Two types of countryside – old hayfields and grazing land – were created through long and continual periods of farming which did not use artificial fertilisers and pesticides. They cannot be farmed within the framework of general agricultural policy, and special preservation measures are necessary if they are not to disappear. The agricultural countryside was originally linked to open types of nature: steppes, beaches, river banks and burned land (Lagerlöf, 1985; Andersson and Appelquist, 1990). When discussing diversity, it is important to distinguish between species which are strongly connected to cultural biotopes and, therefore, dependent on them for continued existence, and species which are connected to rest biotopes, but which also are found in other types of nature. In Norway, it is especially the first type which is threatened; this is a natural consequence of the 'agricultural' countryside being so relatively small and widespread, while 'natural' countryside dominates in terms of area. We find precisely the opposite in, for example, Denmark, Germany and the Netherlands, where the 'original' nature can be found only as rest biotopes in an 'ocean' of cultivated land. The species connected to these 'original' types of nature will be strongly threatened as a result of this strong division (Opdam, 1989).

Agricultural countryside is not a static system, but rather a dynamic agro–ecological system where agricultural practice forms the landscape, its structure and function, and changes the living conditions of animals and plants. Agriculture has changed its character in line with changes in society: the keywords are larger farming units, more specialisation, regional canalisation of production and increased use of external production factors. This has, in its turn, lead to major change in the landscape, and these trends are expected to continue through a further deregulation of agriculture. We shall discuss what repercussions this may have for biological diversity in Chapter 20.

14
Transport

14.1   The impact of transport

As in the agricultural sector, transport has great impact on nature and the environment in Norway. Roughly speaking, we can divide up the sector into the following four sub-groups: international and domestic sea transport, rail transport, roads and air. Table 14.1 shows the contact points between these sectors and environmental and natural resources.

As we can see from Table 4.1, the groups interlock quite closely. The Project for a Sustainable Economy is primarily concerned with the impact on air quality. It is not possible for us to go into more details about the consequences of an ever-more dense road network on bio-diversity because, quite simply, the data is not available. We have divided these effects into three main categories – local air pollution, regional air pollution and global air pollution.

Local air pollution

Local air pollution is characterised by its consequences for the immediate surroundings of the emission source. Transport leads to emission of several local air pollution components. Carbon monoxide (CO) affects the oxidation capacity of the atmosphere and forms with other compounds, including NO_x, ozone under the influence of the sun. It harms crops, coniferous forests and other green plants.

Sulphur dioxide (SO_2) has both local and regional impacts. The greatest effect of the emissions will be in the immediate vicinity of the source. There will, however, be large amounts of water vapour (H_2O) in the air. Some of the sulphur dioxide combines with the water vapour, forming sulphuric acid, which can be transported over great distances

Table 14.1   Transport's points of contact with the environment and natural resources

Various forms of transport	Air quality	Water resources	Land resources	Solid wastes	Noise
Marine and inland water	Local/regional/global pollution	Discharges of chemicals, ballast water, sludge, oil spills Change in natural water systems when building harbours, drainage, etc.	Infrastructure built on land Blocking at closed-down harbours and wharves Use of construction materials (rocks, sand, etc.)	Laid up vessels awaiting breaking up	Around terminals in built-up areas
Rail transport	Pollution (local, regional, global) from diesel locomotives and electricity generation		Tracks, terminals, stations built on land, blocking at closed-down facilities Use of construction materials (rocks, sand, etc.)	Closed-down tracks, scrapped locomotives and rolling stock	Noise, vibration around terminals and along tracks
Road transport	Local, regional, global pollution	Pollution of surface and ground water, change in water systems through road construction	Infrastructure built on land Extraction of non-renewable resources for construction materials	Remaining landfill sites of earth/stone rubble from construction period, scrap vehicles, oil spills	Noise/vibration from vehicles, lorries in towns/along roads
Air transport	Global/regional pollution	Change in water table, river courses and drainage of land areas during construction of airports Chemical runoffs from airport and aircraft operations	Infrastructure built on land Abandoned airstrips and facilities Use of construction materials (rocks, sand, etc.)	Scrap aircraft and waste disposal	Noise around airports and flight paths

Source: UNEP (1993) *The World Environment 1972–1992*, pp. 406–7.

before being precipitated as acid rain. It has been calculated that 90 per cent of the acid rain which falls on Norway comes from sources outside Norway's borders, not least from burning (brown) coal in power stations and households in the UK and Germany. It acidifies fishing lakes, leads to depletion of stocks, and in areas with little chalk in the topsoil, harms trees and plants. In certain cases it can also lead to high levels of algae in freshwater.

Nitrogen oxides (NO_x) contribute significantly to acidification of nature. Reactions between NO_x, CH_4 (methane) and volatile organic compounds (VOC) cause ground-level ozone, which can cause appreciable health problems. NO_x has a significant fertiliser effect and is a consequence of abnormally high concentrations in nature. It can result in high levels of algae in water, which in turn leads to overgrown fishing lakes and depletion of fish stocks. Norwegian emissions of NO_x in Norway were 221 000 tonnes in 1990. They have increased heavily – in direct proportion to the increase in road traffic – but showed a slight decline in 1990 compared with the previous year (Aunan, Scip and Aaheim, 1993).

VOC contribute to the formation of ozone under the influence of sunshine. Ground-level ozone is formed in large amounts during the summer, inhibiting photosynthesis in green plants, even at low concentrations. Plant species exhibit varying sensitivity to exposure, but such an important agricultural crop as wheat has been estimated by researchers to suffer production losses of 30 per cent at an ozone concentration greater than 81 g/m^3 (7-hour mean during the growth season). Ozone concentrations which are higher than 150 g/m^3 have resulted in reduced photosynthesis and needle damage to conifers. It is reasonable to presume that ozone, in conjunction with sulphur dioxide, nitrogen oxide and acid rain is a significant contributory cause of the forest damage we see throughout Europe today. Damage has been proved in certain species of lichen at a much lower ozone level, that is, 59 g/m^3 (SOU, 1993:27 Miljöbalk. Hovudbetänkande). For this reason, the Norwegian Pollution Authority has proposed a limit value for ozone of 50 g/m^3 in the growth season, even though it also points out that it would be desirable to set even lower values as a result of possible combined effects with other pollution components. The recommended limit is exceeded in most parts of Norway, as the so-called background values are in the vicinity of 50–80 g/m^3. Central highly populated agricultural districts in eastern Norway are located in regions suffering from levels approaching the upper limit, and it is possible that crop losses could be appreciable. In Sweden, the annual crop losses resulting from high ozone concentrations are calculated as being as high as SEK 1.4 billion.

Regional and global air pollution

Regional air pollution is characterised by emissions which have consequences for large areas. This pollution can also cross national borders, which means that emissions in one country can harm the environment in another. Both NO_x and SO_2 have major regional impact. Most of the acidification which affects Norwegian watercourses, forests and lakes is a consequence of air pollution coming all the way from the continent and the UK.

Global emissions are characterised by the fact that it is irrelevant where in the world they originate; the impact is the same. CO_2 is the most important of the climate gases because the levels emitted are so high. Norway emits around 34 million tonnes each year, 40 per cent of which comes from mobile sources. When the concentration of the CO_2 content of the atmosphere increases, the so-called 'greenhouse effect' is reinforced, and natural heat loss to space is obstructed. This leads to a gradual increase in the global mean temperature and, according to scientific sources, the world is experiencing the most rapid global temperature increase for 160 000 years. Some scientists also think that NO_x emissions from air traffic could have a deteriorating effect on the stratospheric ozone layer, and also act as a climate gas (affecting the heat equilibrium of the earth). A central point here is ecosystems' tolerance levels. We discuss the environmental costs of transport in Chapter 18.

14.2   Road traffic: we travel more and more

Road transport can be divided into two categories, goods and passenger traffic. We have concentrated on passenger traffic, because goods traffic is not clearly understood and has been only poorly charted. Goods transport is also closely connected to production in individual companies, which makes it more difficult to calculate the consequences of taxes.

The change in Norwegians' travel patterns has been enormous, not least as a result of the growth in private car ownership. While the costs of travel and transport around 1990 amounted to around 19 per cent, or nearly one-fifth, of household consumer expenses, the comparable figure in 1958 was less than 7 per cent (Statistics Norway, *Historical Statistics,* 1994a). Improvements in communication took place long before 1958. In 1945 there was one car per 64 head of population in

Norway, while the figure in 1992 was just less than one car per head. In 1840, 15 000 km of public roads were registered in Norway; this figure increased to 50 000 km in 1958, and by 1992 there were 90 000 km. This is nearly a six fold increase from 1840 to 1992. It is worth taking into account here that there has been significant change in what is regarded as a 'public road', which means that very few of the 1840 public road km would be approved as public roads today. The situation, however, is that the number of road km must be seen in relation to the demand, and that makes it more complicated to draw any conclusions from a comparison.

One fact which tells us more about travel activity is that in 1860 Norwegian railways carried 152 000 passengers over 4.8 million passenger km. Both passenger figures and passenger km increased, reaching their greatest level in the period between 1946 and 1960. Trends changed however, not least because car travel took over from trains. Rail passenger figures fell from 42 million in 1960 to 29 million in 1970, but then recovered somewhat. Mainly short-haul passengers deserted Norwegian state railways, but the fall in passenger km was limited and increased in 1970. The average distance travelled per passenger, which had varied between 23 and 33 km until 1946, increased to 64 km in 1991.

Calculations have been carried out over the total transport services in Norway since 1946. In domestic travel, the figure for passenger km in 1946 was 4591 million. 30 per cent of this was rail transport, 23 per cent private cars. In each of the two nine-year periods between 1952 and 1970 the number of passenger km doubled, until after a 17-year hiatus, in 1987, the figures had doubled again. Growth has stagnated since 1987, and the figures up to 1992 show 52 billion person km each year. In 1992, three-quarters of all travel was private car transport, and less than 5 per cent was rail.

In Chapter 21 we examine more closely how CO_2 taxes can influence trends within passenger transport in relation to the alternative scenarios explained in Chapters 16–18. We will also approach the problem of transport taxation in Chapter 22, where we examine more closely the possibility of higher yields through infrastructure construction.

15

The Impact of More Stringent Environmental Demands on Resource-intensive Industry*

15.1 What can we learn from the experience of Norwegian industry?

Employment effects of environmental demands

A company will maintain its operations as long as current income is greater than current costs (that is, as long as the 'quasi-rent' is positive). This sort of positive quasi-rent will not, however, necessarily mean that a company manages to cover all its costs. It may be that there is not enough left to service its capital – in other words, to reinvest in new capital equipment when the old equipment is worn out, or becomes outdated. In that case, the company would eventually have to close down.

Imagine a company which is forced by the government to reduce its environmentally harmful discharges. If the company does this by only installing purification equipment, without otherwise affecting how the company organises its processes and uses its production factors, then the capital costs will increase. If the purification does not consume operation and maintenance costs then the quasi-rent will remain unchanged. In this case, employment will not be affected as long as the company does not cease operations, although the expected economic lifetime of the company may be shortened as a result of the increase in capital costs. How, and to what extent, this may happen depends on whether the company (a) continues to remain profitable after deduction of all of the fixed costs, (b) continues to manage to meet its variable costs, but not the capital costs or (c) incurs operating losses before

---

* The discussion and results of this and the two next sections are based on preliminary results kindly supplied to the Project for a Sustainable Economy by Rolf Golombek and Arvid Raknerud (1995).

deduction of fixed costs which have arisen after the new environmental demands were introduced. If these demands also involve increased operating costs in addition to capital costs, all of the above will be reinforced and many companies will immediately consider ceasing operations. Even if it is feasible that the increased environmental operating costs result from an increase in labour costs, it is reasonable to assume that the total employment effect would, in this case, be negative.

If, on the other hand, the company employs new technology to adhere to the new requirements, then the impact on employment will be far less clear. In theory, employment may increase as a result of (1) increases in production volumes or (2) the new technology being more labour-intensive than the old (this is probably not a very realistic assumption). It could, however, be possible that the new technology opens new markets for the company – for example, by making it possible to process former waste products into marketable and profitable products. Actual examples of this sort of 'induced technological innovation' are discussed below.

Thus far, the discussion shows that claims and myths that more stringent environmental demands on industry will lead to reduced employment cannot be accepted at face value; the actual links between stringent environmental standards and industrial jobs are quite complicated. In the absence of comprehensive theoretic models for these links, we are looking for the answer to the following two questions, based on panel data from Norwegian industry for the period between 1976 and 1991, using econometric analysis:

1. Is there any statistical connection between employment and environmental demand trends?
2. If the answer to 1 is positive: has the situation been that companies which have had to meet more stringent environmental demands have shown a greater tendency to cease operation, and a lower tendency to increase employment than equivalent companies (size and field) which have not had to satisfy such new environmental demands?

The analysis model is designed so that it should be able to explain any dissimilarities between companies in the terms of market conditions and, at the same time, identify the partial effects of environmental regulation introduced by the National Pollution Control Authority (NPA) on employment and number of businesses which have closed in three different industrial sectors. The analysis concludes that employment and closure effects from the more stringent demands have mainly been insignificant for (1) the chemical industry, but for other sectors like (2) cellulose, paper and cardboard, and for (3) iron, steel and ferro-alloys, the environmentally regulated companies have shown a higher rate of job increase, and have seen fewer closures, with associated loss of jobs, than equivalent companies which have not been regulated. It is, however, important to emphasise that the analytic model was not created as a theory to explain employment trends; all it attempts to do is calculate a statistical relationship between, on the one hand, the frequency of cases of increased employment, reduced employment and company closures, and on the other hand, the extent of the environmental regulation that the company is facing. This analysis does not allow us to come to the conclusion, with any degree of certainty, that more stringent environmental regulations in the period between 1976 and 1991 will reduce the number of company closures or increase employment in the sectors not covered by the analysis.

Possible explanations for observed employment effects

Could the observed development be the fault of poor quality data or even misleading data? The employment figures are from Statistics Norway and the environmental figures are from the National Pollution Control Authority. Both of these sources are regarded as reliable and of high quality. The next question, therefore, is whether the National Pollution Control Authority employs a system demanding more stringent environmental requirements according to the economic viability of the company to bear the costs of the demands. The question may be especially relevant because it is quite clearly cheaper and easier to plan environmental measures when planning a new company than trying to meet new environmental demands in an already existing one. The National Pollution Control Authority does not take into account company profits when imposing environmental demands but, to a certain extent, allows old companies a longer time to adapt, as long as the authority is sure that it will not have harmful effects on health. It is possible, in this way, to reduce adjustment costs, but not to improve profits in comparison to a situation without more stringent demands. Golombek and Raknerud (1995) have tested the econometric hypothesis that the authority should impose more lenient demands when the financial situation of the company is poor, and found no such basis for the assumption.

The question of whether companies have closed when they knew that more stringent demands from the authority were on the way is

more serious. If this were the case then the estimated number of closures of non-regulated companies would be too high. According to the
authority, closures have not taken place after 1975; it is, however, possible that such a situation tipped the balance for several small, elderly
cellulose, paper and cardboard factories which closed down in the early
1970s.

As long as different industrial groups are categoried under the four-
figure ISIC level there will be significant differences between companies within the same ISIC group. The analysis has taken into account
important aspects of this internal sector variation, and what remains
is regarded as having no significance for the conclusions of the
analysis.

If it was the case that consumers of the above industrial products
prefer, and are therefore more willing, to pay more for products manufactured using more environmentally sound processes, it would be
reflected in the price of the companies' products. This is taken into
account in the analysis. It also seems quite clear that, to the extent that
it is a significant factor, the manufacturers take it into account in their
choice of manufacturing processes and purification equipment. Several
manufacturers of cellulose and paper have even introduced more stringent environmental standards than those set by the authority in order
to satisfy the demands of foreign customers.

Perhaps the most interesting explanation of the observed trend will
prove to be connected with the fact that more stringent environmental
demands extend the information base which company management
draw from when they decide new investment and select production
and purification processes, marketing strategies and internal company
organisation matters. Is it possible that, for example, the National
Pollution Control Authority provides the companies with new information which management would not otherwise have gained, or
secured, access to? One example could apply to information about
technology which can convert waste products into marketable goods.
Because the National Pollution Control Authority works in partnership
with sister authorities in other countries, it has access to a lot of technological and market information. The information includes knowledge of potential environmental regulation in other countries which
individual private companies might not have learned about on their
own. It is then possible to speculate that the authority, when imposing
demands which meet future environmental demands in international
markets, contributes to improvement of the competitive edge of the
Norwegian exporters which have to meet its demands. The authority

itself claims that it plays a part in the distribution of such knowledge, not least to small and medium-sized businesses (SMEs).

The other condition which should be noted in connection with the increased environmental demands is that they force companies to review many of their production processes, their administration and their logistics. This sort of new information can reveal waste of resources and result in changes which increase the company's productivity and profits. On the other hand, environmental demands which result only in installation of purification equipment merely increase costs, and therefore should be regarded as 'non-productive' by the company concerned. The net effect can be measured only empirically. To enable this, the total factor productivity (TFP) is calculated as an expression of the difference between relative growth in production and production factors for the relevant companies in the survey. Golombek and Raknerud (1995) found, in the cellulose, paper and cardboard sector, and in the chemical industry, that the TFP was somewhat higher for the non-regulated than for the companies on which environmental demands were imposed. For the iron, steel and ferro-alloy industry, however, the TFP was somewhat higher in the environmentally regulated companies, making it difficult to draw any general conclusion on the basis of the available data. It also proved, under closer scrutiny, that there was no systematic difference in the mean investment per employee within each of the three ISIC sectors.

15.2   How major Norwegian companies have adapted to environmental regulation

It is not possible to establish any robust theory for the synergy effects of environmental demands from the above analysis. It may, however, be of interest to consider actual examples from recent Norwegian industrial history which indicate that environmental demands do have such a synergy effect.

Borregaard Cellulosefabrikk, a cellulose factory on the Glomma river in eastern Norway, caused significant pollution of the river and Oslo Fjord, into which it runs. The company was required to invest significantly in environmental measures. Two major by-products from Borregaard's acid–sulphite production process are hemicellulose and lignine. These originate from components in the wood pulp which is not used in refined cellulose. Even before the new requirement for reduced discharges came, Borregaard were aware that it was possible to produce ethanol and animal feeds from hemicellulose. One result of the discharge requirements was that Borregaard decided to convert derivatives from the hemicellulose fraction (sugar) to industrial alcohol (ethanol), building one of Scandinavia's largest ethanol plants. Then they discovered that it was possible to convert the lignine component into vanilla essence (vanillin), which is an important raw material in the production of ice cream, chocolate and cosmetics. Following these changes, the organic discharges from the factory are now only a tiny fraction of its raw material consumption. Borregaard has also become the leading global manufacturer of vanillin and one of the leading manufacturers of lignosulfonate.

Norsk Hydro's fertiliser production at Herøya is another example. Huge quantities of brown smoke, consisting of nitrous oxide from nitrate production, were a source of great irritation for surrounding neighbourhoods. In the early 1980s, Norway signed an international agreement to reduce emissions of nitrous oxides and, as a result, attention was directed at Norsk Hydro's nitrate production. As a result of environmental demands the company decided that it could install equipment which retained the brown, nitrous gases and use them as a factor in the production of nitrate fertilisers. It proved that the investment in recycling equipment allowed the company to produce more fertiliser from the equivalent amount of raw materials, at the same time as emissions from factory smokestacks took on a far less provoking colour and were reduced by 90 per cent.

The third example is from Elkem's silicon production, which is based on heating quartz to extremely high temperatures in the presence of coal. The temperatures are so high that the some of quartz vapourises and later condenses into small particles which irritate the lungs. The company was required to invest in equipment to reduce these emissions. The available equipment was expensive to acquire and had high running costs. This gave the company the incentive to research alternative solutions which could put these small irritating particles to productive use. As a result of the requirements, it was found that when these particles were added to concrete they increased its tensile strength immensely. This opened a completely new market for concrete constructions with specially high requirements for tensile strength, including drilling rigs, because the new technology allows use of significantly less concrete while still retaining the strength of the rig. The product also proved to exhibit special beneficial qualities as an additive in the plastics and paper industries.

What about situations where double dividends cannot be assumed?[1]

When approaching many issues from a partial perspective, it can often be useful to assume that double dividends cannot be quite simply expected. Examples could include the direct effects on employment of increased use of certain Norwegian natural resources. We have focused especially on two types of resource use, (a) new hydropower schemes and extension and upgrading of existing hydropower schemes, (b) extraction and use of natural gas for production of electrical power in Norway. One actual example of each of the two types of project is discussed: For type (b), the Heidrun project on the Halten Bank, which will pipe gas to the mainland with a potential gas plant at Tjeldbergodden, was chosen. For type (a), there are to be new hydropower schemes, and extension and upgrading of the existing scheme at Sauda, in relation to the main alternative proposed by Statkraft. The study examines the impact on employment of the project itself, and operation of the scheme. Most of the discussion will concern the immediate use of labour in building and operating the schemes, and supplying them. We will also mention, although somewhat sketchily, the impact on total local and national levels of employment which results from this use. Any further effects on the economy, however (through changes in demand or economic policy), which result from the income effects of the project on the community, are ignored.

For the gas extraction and use projects, it should be quite clear that increased extraction and Norwegian use of the gas will lead to increases in labour needs and therefore (at least potentially) to an increase in employment. Because increased global use of natural gas leads to increased carbon dioxide emissions, contributing to a serious global environmental problem (as far as it is not possible to prove that increased use of natural gas replaces coal and oil to such an extent that the total CO_2 emissions are lower than they would have been had the supply of natural gas not been increased), there would seem to be a clear conflict here between employment and environmental considerations, at least where the gas production itself is concerned (as long as this leads to an equivalent increase in the global use of gas). As far as national vs the non-national use of gas resources is concerned (the national case is production of electric power from a Norwegian gas-fired power station) there could, in principle, be thought to be fewer extra environmental problems connected to national rather than foreign use, especially if this use, in itself, does not lead to greater global use of energy than that which would have resulted from foreign use. It is, however, quite possible that national use of the gas would lead to less effective energy production on a global scale, especially since gas used in some other place would replace energy produced at higher cost, or less efficiently. It may also be that gas used in Norway to produce energy for export would mean greater transfer losses than the equivalent losses which result from direct gas export. Some local environmental problems would arise from national use; these are explained in ENCO (1993).

The conflict between environmental and employment considerations is equally clear in the case of hydropower as for gas, at least as far as the development and extensions projects are concerned. Based on the analyses behind the report, *Norwegian* (Collective Plan for Watercourses) 1990, it is quite clear that all watercourses in which a new project, or extension of an existing project, are feasible would suffer environmental strain from the project. The environmental strain which results from extension projects could probably be less, perhaps even non-existent. On the other hand, it is somewhat difficult to imagine hydropower upgrade schemes which would result in direct environmental improvement. All in all, the conflict is present for most potential hydropower projects (whether they are pure new construction, or combinations of the three types above).

The analysis is very partial in the sense that it addresses only the employment impacts of the use of natural resources, and only to a limited degree addresses the problem of what sort of environmental strain this use of natural resources inflicts on Norway, or how the loss of natural resources should otherwise be measured. These questions are, of course, essential in the assessment of total socio–economic profitability of projects for extraction and use of the relevant natural resources. The main point here is, however, that such analyses must be made separately, independently of the analyses presented here.

It is also important to be quite clear that the alternative to use of Norwegian natural gas which is proposed – that is, generation of electrical power in Norway – is only one of several possible alternatives. In addition to direct export of the gas, another very realistic alternative would be use of the gas as a production factor in the Norwegian petro-chemical industry, for example in production of methanol. We have not discussed this possible use here.

Part IV

Norway in 2030 – Can the Economy Survive with 'Green' Policies?

16
What Determines Economic Development in Norway?*

In a long-term perspective, economic growth is determined by technological development (including increases in the knowledge base) and access to resources in the economy. This includes population growth, growth in man-made capital and natural resources. Norway exports much of its production and imports a lot of its consumer goods. Conditions in the international economy are, therefore, also important for the course of the Norwegian economy. In this chapter, we shall briefly summarise some of the important preconditions for the reference course for the macroeconomic model scenarios. The preconditions are mainly the same as in the last Long-Term Programme (LTP) (White Paper, 4, 1992–3). Later we shall see what each of the alternative preconditions could mean for economic growth and the environmental impact of such growth. A number of these preconditions, however, will be identical in all of the alternative scenarios.

16.1 Population growth and the workforce

Population growth

The analyses are based on the population growth shown in Figure 16.1, which also shows the actual population growth in the period 1970–90. It can be seen that growth is expected to be slight throughout the entire simulation period. In 2030 the size of the population is expected to slightly exceed 4.7 million, compared to 4.3 million in 1990.

---

* This chapter builds on Alfsen, Larsen and Vennemo (1995).

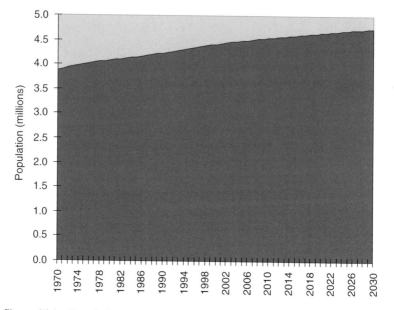

Figure 16.1   Population growth (million)
Source: Alfsen, Larsen and Vennemo (1995).

Age profile

The age profile of the population is shown in Figure 16.2. This shows the actual age profile in the period 1975–90 and also the predicted age profile in the future. The main characteristic is the proportion of elderly citizens is expected to increase as we approach 2030. Even so, the number of people of working age increases from 2.5 million to 2.7 million over the projection period.

Disability

There is great uncertainty connected to the trends with respect to future disability levels. The constant disability access rates which were observed in 1990–1 have been used as a basis for calculation. We assume, as in the LTP's reference alternative, that the health and social sector does not improve productivity with respect to reducing disability levels, which would otherwise have led to a gradual reduction in these access rates. This leads to an increase in the number of disabled from 239 000 in 1990 to around 370 000 in 2030.

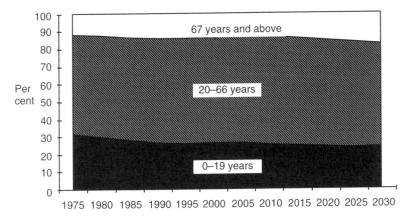

Figure 16.2   Population age profile (per cent)
Source: Alfsen, Larsen and Vennemo (1995).

Working hours

It is presumed that the working hours per employee fall by 9 per cent from 1990 to 2010, and are thereafter constant. This is equivalent to a reduction in the mean number of hours worked per employee per year from 1400 to 1300. This is far less of a reduction than we have seen during the last decades. On the basis of this, and what we have previously said about growth in the workforce, it can be seen that the total number of hours worked is expected to increase by 5 per cent from 1990 to 2010. From 2010 to 2030, the number of hours worked is expects to fall by around 1 per cent.

16.2   The petroleum sector

Crude oil prices on the world market and USD exchange rates are very significant for the Norwegian economy. The calculations are based on a constant crude oil price (measured in fixed prices) of NOK 130 per barrel to the end of the projection period. A significant increase in petroleum production, from around 132 million tonnes oil equivalent (mtoe) in 1992 to 165 mtoe in 2000 is assumed. Thereafter, production decreases to 137 mtoe in 2010 and 87 mtoe in 2030. Production of natural gas will increase steadily to around 60 mtoe during the 1990s, and it is assumed that the level will remain constant throughout the calculation period.

The Long-Term Programme (1994–7) (Ministry of Finance, 1993) examines in detail what an increase in real terms of the price of crude oil would mean for future growth in the Norwegian economy. The significance of the petroleum industry for the Norwegian economy has also been studied more closely by Bye *et al.* (1994). Later, we shall examine the significance of reduced levels of oil and gas production in alternative long-term scenarios.

16.3   The public sector

The development of the public sector is based on the demand that the budget, over the forecast period, must balance. This means that there is no room for increase in state real investment from 2000. The financial holdings of the public sector are set to decrease until 2000 and then increase somewhat in the years until the fall in oil production and the entry of the baby boomers into the ranks of senior citizens. For this reason, it is presumed that the public sector must show a modest budgetary surplus in 2010 and a minor deficit in 2030.

Subsidies to industry will, in real terms, be reduced by 40 per cent from 1992 to 2010. The total costs of old age and disability pensions will, however, grow by a mean 1.7 per cent per annum over the same period and by 1.8 per cent subsequently. This is also a function of population trends, and especially the number of people drawing disability pensions. The LTP examines especially the effect of alternative assumptions about the number of disability pensioners on economic trends.

16.4   Technological change: what affects technological advance?

The growth in the total factor productivity (TFP), for simplicity's sake called technical or technological change, is that part of the GNP growth which cannot be traced back to increased use of labour or capital. The rate of presumed technical change in the transport sector is relatively high, as a result of improved energy efficiency. The growth in the factor productivity, in a total view during the period 1989–2030 is assumed to be 80 per cent, or a little over 1 per cent per year. This is appreciably lower than was observed in the 1970s, although somewhat more than the change in 1989. (For a more detailed description of the product function and technical change, see the Project for a

Sustainable Economy, *Report* 15, 1995.) Part of the reduction in the historic rate of technical change is attributable to the change from industrial production (where technical change has traditionally been rapid), to heavier provision of services, which shows slower technical change. Certain industrial sectors have exhibited slower technical change through the 1980s than they had during the 1970s.

Technological change, as modelled here, allows increased production without increased use of production factors. Emissions to the air in the model are, for example, determined by production factors: increased production which results from advances in technology will not, therefore, lead to increased environmental strain. The rates of technical change which are used in our model calculations have been, on average, set equal to what we have observed historically (lower than the 1970s, higher than the 1980s) – in other words, we presume that the same forces which have led to technical change earlier will also apply in the future. These forces have partly manifested themselves through technical innovation (inventions, and so on), improved organisation of production (perhaps as a result of higher levels of education) and by the nature of produced goods changing over time. When our projections are based on the historic rates of change, this is more an expression of our lack of knowledge of what actually drives the whole picture of technical advance, rather than a belief that technical change will take place at an unchanged rate of growth in the future. It is quite possible that increased consciousness about environmental problems will, for example, lead to more environmentally sound production in the future, although this has not been quantified in our models. In the same way, it is also possible that policy changes – for example, increased concentration on environmental taxation – may affect the rate of technical change. This may take place through environmentally sound technology being chosen when other, more polluting solutions are more expensive. Policy change, when introduced on a large scale, influences technical developers in, for example, the automotive industry, so that greener alternatives become available. These conditions are not included in the analyses as presented here, and this must be taken into account when the results are interpreted.

16.5 The international economy

A growth in the GNP of most prominent trading partners of 2.5 per cent per annum was expected up to the turn of the century. Growth thereafter is now expected to be somewhat slower, at around 2 per cent per year. The reason for the lower growth is primarily slower population growth, and therefore, slower growth in the workforce. In comparison, it is worth noting that GNP growth was around 2.2 per cent in the period 1973–90. Foreign trade has traditionally increased more quickly than GNP; for this reason, we presume a market growth for Norwegian export products of around 4 per cent per annum until 2000, and thereafter a gradual decrease to an annual growth rate of around 2.5 per cent per annum.

16.6   Environmental taxation and petroleum wealth

The shares of the various components of a country's wealth can change significantly over time, partly as a result of the disposition of wealth components, partly as a result of external circumstances which affect the components' value. One example of the first type of change could be that Norwegian petroleum wealth (capital) will be invested in infrastructure, structures, production equipment (real capital), increased skills through better education and training (human capital) or foreign claims (the establishment of a petroleum fund). In this way, it is reasonable to assume that over a 40–50-year period most petroleum wealth will be invested (or consumed) so that the real capital holdings, the knowledge capital and the net foreign claims have increased.

One example of external effects influencing the extent and formation of national wealth would be international agreement on the introduction of a significant CO_2 tax. The effect of such a tax on wealth would depend on the background for its introduction, even if it is clear that the tax would reduce the share of the oil and gas price due to the producers. If it is thought that there is a basis for claiming the consumption of petroleum and coal leads to unwanted climate change to such an extent that the costs of any given reduction of CO_2 emissions are less than the environmental, health and production losses involved in continuing the trend of energy use, then the CO_2 tax should be regarded as a pure environmental tax which will internalise the negative environmental externalities of fossil energy use.

If the necessary level of CO_2 taxation in this case is so high that production of oil and gas in the Norwegian sector of the Continental Shelf ceases, the conclusion would be that Norway actually has no petroleum wealth. Only market imperfections have made it possible for Norway to extract an income from petroleum production. The negative environmental externalities of petroleum consumption are distributed world-wide through increases in CO_2 concentrations and the climatic effects resulting from them.

If, on the other hand, the potential global climate changes are not considered to be likely, and an international or global CO_2 tax is still introduced, then the reason must be sought in fiscal policy. In that case, Norway primarily has a petroleum wealth, although it may be reduced or eliminated as a result of the introduction of the international agreement.

The more inflexible the petroleum supply is, the greater the reduction in producer income will be, and the consumer price will only be marginally increased. In this case, there will be few global environmental gains from the taxation, while a high-cost petroleum-exporting country like Norway will lose a lot, if not all, of its petroleum wealth. Calculations from Statistics Norway (Kverndokk and Rosendahl 1995) indicate that a global CO_2 tax of $3 per barrel would reduce Norway's petroleum wealth by 15–22 per cent, while a tax of $10 per barrel would lead to a reduction of 47–68 per cent, and a tax of $20 per barrel would virtually eliminate Norwegian petroleum wealth.

If Norway were to take the initiative for reductions in global CO_2 emissions and introduce a unilateral Norwegian CO_2 tax increase, which no other country follows, the effect would be quite different. This taxation would have an impact only on Norwegian petroleum consumption, and through it Norwegian production and consumption, while Norwegian petroleum export would be unaffected. In this way most of the petroleum wealth could be distributed in time with petroleum production.

It is, however, important to take notice that it is human capital and technological change which are the dominating components of national wealth in the Norwegian economy, and which drive productivity trends providing the engine for growth. The growth in returns on the real capital and petroleum wealth can explain no more than nearly 30 per cent of the growth in disposable real income *per capita* according to the state LTP.

17
The Analysis Framework of the Alternative Scenarios[*]

17.1 Introduction

The alternative scenarios which we present here are at the core of the Project for a Sustainable Economy. We shall attempt to sketch alternative environmentally driven courses of development for Norway using the macro-economic planning models which are regularly used for long-term planning by the Norwegian government. The difference between the calculations we shall carry out, and the perspective analyses up to 2030 of the government Long-Term Programme (LTP) 1994–7, is that we explicitly take on environmentally loaded development targets. We also determine the value of many of the exogenous variables based on the demands handed down by the 'Environmental Parliament'. On the basis of the preconditions in the models and our estimations, the model will calculate trends in total production both in various industries and in the private and public sectors. The models also describe Norwegian energy production and consumption. Emissions and discharges of various pollutants from fossil fuels and a range of industrial processes are calculated in a supplementary model. The models thus allow us to assess economic, energy consumption and certain environmental trends in their context. This means that we can directly compare our results with the results of the LTP.

We will also compare the results with the demands on endogenous variables as prescribed by our 'Environmental Parliament' on the basis of the requirement for sustainability, as set out in Chapter 10. By comparing with the LTP, we can test the economic consequences of imposing explicit ecological constraints on the economy. In addition, we can

---

[*] This chapter builds on Alfsen, Larsen and Vennemo (1995).

also test how far the economic planning models can be 'stretched' in an ecologically sustainable direction. In the rest of this chapter, we will take a brief look at the models, before finally attempting to indicate the frameworks of the various scenarios.

The model simulations which are reported here are built on two macro-economic models developed in Statistics Norway's research department. Most of the model simulations have been carried out using the long-term equilibrium model MSG-5. This is the same model used by the Ministry of Finance when carrying out the long-term perspective calculations in the latest LTP (Finansdepartementet, 1993). We have chosen to base the analyses we present here on the so-called 'Base case' in the LTP.

Most short-term phenomena and problems are difficult to illustrate when using a long-term general equilibrium model. For this reason, the medium-to-long-term (5–10 year) model, MODAG, is used to illustrate some of the transient problems which can occur when adapting the economy to meet the demands which we have listed above. To understand the limitations which are inherent in the model apparatus, and thereby give a better interpretation of the model results, it is important to know a little about how the models are constructed. This chapter describes MSG-5 in most detail, but also considers the most important differences between this model and the disequilibrium model MODAG.

17.2   The time perspective

Based on the overall sustainability perspective, it is appropriate to express the environmental demands to the models for the project as targets to be reached by 2030, which is also the time frame of the government's LTP. This makes it most appropriate to select the so-called MSG model for analyses of alternative future perspectives. It is, however, of some interest to examine the problems involved in the transition to a new, long-term state of equilibrium. The MODAG model is used for this in the first environmental scenario alternative (that of reduced oil and gas production) which illustrates the nature of the problems the changes can cause in the period of flux.

17.3   A brief description of the models

MSG is a long-term model – that is, it has a time perspective of 15–40 years. This model has been used, and indeed still is used, by the Norwegian Ministry of Finance for analysis of long-term trends in the domestic economy. It is important to note that this means that we apply the same analytic tool in our analyses as the ministry and government use in theirs.

MSG is a multi-sector equilibrium model for the Norwegian economy. This means that the supply and demand for each precondition are the same in all markets, and consumers and producers exploit all available resources. All of the available labour is used and labour supply is exogenously determined – that is, it is determined from outside the model. The model does not distinguish between different geographical localities in Norway for the activity.

The economic growth is determined to a high degree by technological change, growth in real capital and job supply, and access to raw materials and natural resources. Technological change is historically calculated: this means that if the model is to calculate a zero-growth situation, then a downturn in resource access would have to be postulated.

The model is built up of around 30 different sectors, which in their turn consist of various products. The model is a so-called input–output model, which means that products from one sector can be production factors in several other sectors, which in turn can be production factors in other sectors again. This means that, in practice, if prices rise in one sector, it will affect not only that sector but also others. (For more detailed description of the models and the detailed sector breakdown in the analysis, see PBØ, *Report* 15 August, 1995.)

The performance of the model is determined by prices and income. This means that changes in beliefs which are reflected in the behaviour of individuals are not indicated by the model. Consumption of a product can be reduced in two ways:

1. The price increases so that it becomes more expensive to acquire the product.
2. Incomes are reduced so that the purchasing power of the individual with respect to this product is reduced. In the case of a price increase we also have an income effect, which means that the real income is reduced when one or more prices rise.

MSG provides a detailed description of the production and use of energy in Norway. Using supplementary models, it is possible to forecast the emission of various pollutants from use of fossil fuels. The model can then be used to assess the interlinkages between trends in the economy and certain environmental conditions.

MODAG is a medium-to-long-term model which also sees the nation as one. It has a time frame of between 5 and 15 years. Virtually all of the parameters in the various sub-models are econometrically estimated and based on the national income accounts, and the model is regularly updated.

The sub-model for household behaviour includes labour supply. It does not require full exploitation of resources. This means in practice that it can allow for unemployment.

The size of the economy is determined by the access to resources, resource exploitation, balance of trade and, to a limited extent, productivity. Economic growth is determined by growth in resource access, changes in resource exploitation, the balance of trade and balance of the budget.

Performance is determined by prices, income and trends. Trends may reflect gradual changes in attitude which depend on prices and incomes.

17.4   Closing the models

A macro-economic model can be 'closed' in various ways, and the method of closure will affect the characteristics of the model. We can explain closure in this way: *A macroeconomic model may have a formula which determines private real investment – for example, that saving is a constant fraction of income.* In this way, it is possible to have a formula which determines private real investment – for example, that investment is dependent on the interest rate in the economy. If the model is going to be used on behalf of the government, it would be natural to make public saving and real investment 'exogenous' – in other words under the direct control of the user. In the final analysis, we would also like to make the current account exogenous because this is a quantity which the model user has wishes and opinions about. The current account is equal to a country's financial investment.

So, we can either denote private and public saving, private and public real investment and the country's financial investment as an equation or an exogenous variable. One of these must, however, give way for an economic definition which we are compelled to allow for: The aggregate of private and public savings must equal private and public real investment *plus* the financial investment of the country. Ignoring this definition will lead to logical errors in the model. When a definition is allowed to determine a variable, this is called closing the model. We can choose which variable we want to allow the definition to determine, and different choices lead to different ways of closing the model.

In the simulations which are presented here, the MSG model is closed by letting the definition determine private savings. As we have said, this affects the characteristics of the model. For example, in one of the alternative scenarios, which we shall present later, it is presumed that investment in oil and gas fields is reduced. The model interprets this as a reduction in total investment, because other investment is determined by interest rates. Reduced total investment then leads to reduced total savings, which means reduced private saving because public saving is exogenous. The size of private incomes is unchanged at first. Reduced private saving will make room for increased private consumption. All in all, the resources that the economy saves in reduced field development are withdrawn as increased private consumption (in time the reduced production will, however, lead to a stagnation in private consumption).

How would it be if we allowed the definition to determine private real investment? In that case, lower field development would provide an increase in private onshore investment, because private saving would be constant as long as private income was constant. The total level of investment would, therefore, be the same as before, and private onshore investment would have to increase to compensate for lower investment in oil and gas fields. The one closure leads to increased consumption, the other to increased onshore investment.

It is not *a priori* possible to say that one way of closing a model is better than another. On the one hand it is reasonable that domestic investment takes place on the basis of which profitable investment objects are present, not on the basis of how much is invested in oil and gas fields. This supports the closure which is actually chosen for MSG. On the other hand, it would seem to be reasonable that some of the loss of investment in the fields in the alternative scenario is compensated for by increased investment on shore. This does not support the closure actually chosen, but does support the other alternative. One advantage of the solution actually chosen is that changes in private consumption can be interpreted, to a certain extent, as an indicator of the costs (or dividends) of the chosen policy, because many of the effects are manifested as changes in private consumption. We will return to this in Chapter 18. A simulation has also been carried out using the medium-to-long-term model MODAG, which is a non-equilibrium model which models all demand. In the case of any shock in the economy, wages will not manage the labour market, possibly resulting in unemployment; in the model, private investment is determined by operating results and production. Private saving is determined on the basis of the consumption functions. Public sector saving is determined as the difference between tax income and exogenous expenses. Finally, the current account follows residually, that is, closes.

18
Comparison of Governmental Projections with Environmentally Adjusted Projections*

We shall now examine more closely how the indicators derived from the 'Environmental Parliament' (see Tables 10.1 and 10.2, pp. 106–7) affect long-term trends (projections) in comparison with the trends shown in the government's Long-term Programme (LTP). The computations are carried out using the 'MSG-5' model. MSG is a relatively disaggregated model both for sectors and goods, although not all indicators can be illuminated using the variables of the model. Indicators which treat polluting discharges and emissions can be studied because these are explicitly calculated in a dedicated model. Another subsequent model calculates a number of indicators, including consumption of petrol and other fuel, heating oil, and so on, measured in physical units. The actual consumption of coal, coke, wood, district heating, LPG and other gas is, however, not explicitly projected in the model. The indicator for energy consumption calculated using MSG will not, therefore, take into account these energy sources, although they will be included in the general product range in each sector.

18.1   The reference path

The LTP's 'basic alternative'
The reference path is mainly equivalent to the 'basic alternative' in the LTP (Finansdepartementet, 1993). Table 18.1 shows the trends in some key macro-economic indicators in the reference path, together with historic figures.

---

* This chapter builds on Alfsen, Larsen and Vennemo (1995).

Table 18.1 Trends in some key macro-economic indicators, 1970–2030 (annual percentage growth in fixed prices, level figures, 1989 NOK 1 000 million, reference alternative)

	Annual growth	Level	Annual growth	Level		Change	
	1970–80 (%)	1980–1989 (%)	1980–1989	1989–2010 (%)	2010–30 (%)	2030	1989–2030
Disposable real income	4.0	1.3	499.7	2.3	1.2	1029.2	106
GNP	4.7	2.5	623.0	1.8	1.1	1134.4	82
Mainland	3.7	1.6	516.0	2.0	1.4	1049.9	104
Norway industry and mining	1.5	0.3	87.0	1.6	1.6	179.7	107
Exports	5.4	5.0	261.5	2.1	1.3	555.7	113
Imports	3.4	2.4	234.7	2.9	1.6	585.4	149
Private consumption	3.6	1.8	312.0	2.5	1.7	729.9	134
Public consumption	5.3	3.2	131.1	1.6	0.4	196.1	50

Source: Alfsen, Larsen and Vennemo (1995).

The calculations indicate that prospective growth for the Norwegian economy is lower in the long term, compared to historic trends and estimated trends abroad. This is linked to labour supply, decreasing activity in the oil and gas industry and declining growth impulses from other raw material-based industries. The GNP for mainland Norway grows by around 2 per cent per year until 2010, then declines to around 1.5 per cent per year. In contrast to the historical trend, industrial activity increases roughly in pace with GNP after 2010. This can only be interpreted as placing the demand of meeting the target of full employment and a strong foreign economy on the shoulders of industry. Job growth will, however, primarily take place in service industries. This is because these are more labour-intensive than manufacturing industry and that technical progress is generally lower in the service sector than in industry.

Even if growth rates decline, Norway in 2030 will be considerably more affluent than present-day Norway. We will return to the question as to whether this increased affluence will impose more strain on natural resources and the environment.

The share of the gross production of primary industries is maintained at around the 1989 level in the reference path (agriculture is reduced, while forestry increases). Several production sectors decline in relative size in the projection period. This applies to manufacture of consumer goods, metals and the construction sector. The greatest reduction comes, however, in oil and gas production, where the share of the total gross production is reduced from around 10 per cent in 1989 to around 4 per cent in 2030. Such sectors as wood-processing, machine and equipment manufacture, private services, trade and transport grow in significance in the reference path. Table 18.2 shows trends in the various consumption patterns in the reference path.

The strongest growth is shown in 'holidays abroad' and in such hard consumer goods as cars and furniture. The growth in energy use for household heating is, however, low. The same applies to increased petrol consumption, and the reference path allows for great technological advances in the automobile fleet (improvements in fuel efficiency, 2.2 per cent per annum).

Emissions to the atmosphere in the reference path

The premises as to atmospheric emissions from the 'Environmental Parliament' are linked to reduction in CO_2, CH_4, N_2O, NO_x, SO_2, and non-methane volatile organic components' (NMVOC) emissions. Figure 18.1 shows emission trends in the reference path. None of the emission reduction demands are met in the reference path.

Table 18.2 Private consumption, 1989 and 2030 in the reference path (1989 NOK 1 000 million)

Consumer goods	1989	2030	% growth
Electricity	16	22	35
Oil for fixed installations	2	3	27
Petrol	15	18	22
Car purchases	11	36	239
Public transport	15	36	142
Food	58	97	67
Beverages and tobacco	22	48	124
Other goods	32	79	151
Clothes and shoes	22	58	161
Furniture	19	59	203
Housing	40	81	101
Other services	36	110	206
Holidays abroad	20	75	270
Total consumption	**312**	**730**	**134**

Source: Alfsen, Larsen and Vennemo (1995).

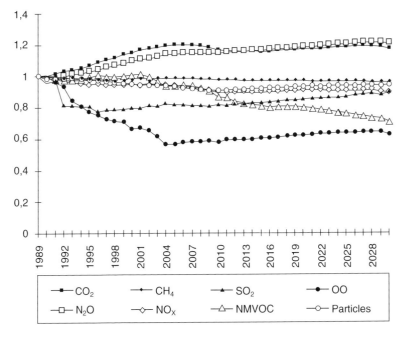

Figure 18.1 Atmospheric emissions in the reference path (indexes: 1989 = 1)

By 2030, SO_2 emissions will be reduced by 63 per cent compared with 1980. The demand is for a 90 per cent reduction, so it is nearly met. The demand on NMVOC reduction is also nearly met, as the reduction in the reference path is around 40 per cent. The demand for a 60 per cent reduction in greenhouse gases is, on the other hand, far from met, as there is in fact an increase of around 20 per cent in these emissions in the reference path. The goal of a 50 per cent reduction in NO_x emissions is also far from being met.

Even if the goals are not achieved, the emission increase is far lower than the growth in economic activity. When looked at in this way, the reference alternative can be said to provide a picture of a society which is, relatively speaking, more environmentally caring than today. If we are to approach the goals, either the composition of economic growth must be radically changed, and/or the rate of growth must be reduced.

Production of metals, chemical products, production materials and timber in the reference path

Production of crude aluminium and fertiliser should, according to the demands of the 'Environmental Parliament' presented in Table 10.1 not exceed present-day levels. In the case of cement production, the demand is for a 70 per cent reduction. In MSG-5, the production of crude aluminium is included in the 'Metals' sector, and production of fertiliser is in the 'Chemical products' sector. Cement production is included in the 'Production factors' sector. If we are to be able to say anything about trends in production of these goods, we must presume that all goods included in the product aggregate grow in time with each other. The growth in production of crude aluminium is then assumed to be equal to the growth in the metal sector, and so on. In the reference path, gross production in the metals and chemical industry sectors increases by 35 per cent and 45 per cent respectively between 1989 and 2030 (see Figure 18.2). The demand for unchanged production levels in crude aluminium and artificial fertiliser are, therefore, not met in the reference path.

Cement production *per capita* must be reduced by 70 per cent by 2030. Population increases by 12 per cent from 1989 to 2030, while cement production increases by 55 per cent. This means that there is an increase in per capita cement production of more than 40 per cent. Gross production in forestry increases by 180 per cent from 1989 to 2030 in the reference path – that is, the demand for unchanged production levels for timber is far from being met.

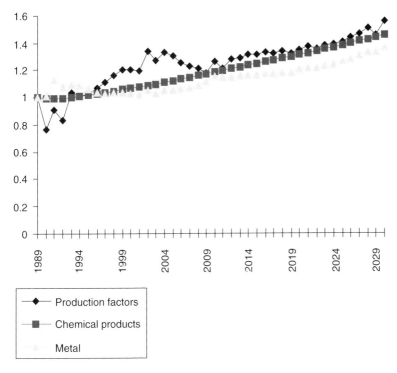

Figure 18.2   Gross production in the metal, chemical products and production factors sectors (indexes: 1989 = 1 in the reference path)

Consumption of cement, metals, timber and fertilisers in the reference path

Per capita consumption of cement, steel, aluminium, lead, copper and fertiliser must be reduced by 80 per cent by 2030 in order to meet the demands of the 'Environmental Parliament'. There are no separate variables for such consumption in the model. Domestic consumption of cars, housing and furniture/electrical equipment can, however, indicate something about the consumption of metals, timber and cement. There is a heavy increase in consumption of cars, housing and furniture in the reference path (239, 101 and 203 per cent, respectively from 1989 to 2030). In comparison, population growth is only 12 per cent, so these demands are far from being met in the reference path. Another indication of the trends in consumption of cement and timber could be the trends in the building and construction sector. If this is taken as an indicator, the production of cement and timber is doubled between 1989 and 2030 in the reference path.

If the total domestic consumption per head of these products is examined (that is, Norwegian production *plus* import, *minus* export), cement consumption increases by 91 per cent, metals increases by 46 per cent and fertiliser increases by 30 per cent.

Hydropower, atomic power and gas power

The demand regarding atomic power is met in the reference path, since no such power is generated. The demand that no gas-fuelled power shall be generated is not fully met in the reference path, since in 2030 2.1 TWh of electricity is generated in a gas-fired power station. This is, however, only 1.5 per cent of the total power supply. The LTP points out that there is a great deal of uncertainty around the profitability of gas-fired power stations, since the difference in costs between hydropower and gas power are both very small and uncertain towards the end of the calculation period. Total electricity generation in 2030 is 140 TWh, – that is, the entire hydropower potential of the National Collective Plan is exploited.

Development of oil and gas fields

The demand regarding development of oil and gas fields states that they should not be developed north of the 66th Parallel, in the Skagerrak and closer to the shore than two days drift time. Nor should new, purely oil-producing fields be opened. These demands cannot be said to be met in the reference path, since such limitations in oil and gas activities will not be applied as long as the fields are profitable.

Energy consumption in the reference path

Economic growth moves in the direction of increased energy consumption, while technical change means that some of the growth can take place without any such increase. Even so, the total domestic energy consumption *per capita* increases by 16 per cent from 1989 to 2030 in the reference path. Total energy consumption includes consumption of oil and so on, and electricity in the manufacturing sector and households (coal, coke, wood, district heating, LPG and other gas is not included[1]). The growth in total oil consumption for transport purposes is 43 per cent from 1989 to 2030, while the growth in oil for heating purposes is 28 per cent. Electricity consumption also increases by 28 per cent. This assumes that power-intensive industries (chemicals and metals) are allowed to retain power contracts of 30 TWh (equivalent to about 1990 consumption) at the same real price as today, and that it will not be profitable for the industry to purchase electricity above this at market prices.

18.2 Alternative future scenarios

About the scenarios

PSE's alternative-future scenarios include the following:

Alternative 1 The significance of reducing oil and gas production in especially vulnerable areas.
Alternative 2 The significance of a high CO_2 tax.
Alternative 3 The effect of combining a high CO_2 tax with reduced oil and gas production.
Alternative 4 The significance of exempting certain industries from the CO_2 tax.
Alternative 5 The significance of reduced working hours.

Alternative 1: reduced oil and gas production

The philosophy behind alternative 1 is to examine the consequences of avoiding the development of any viable oil and gas fields in vulnerable deep-sea and coastal waters. The motivation of the 'Environmental Parliament' is partly that very vulnerable ecosystems should not be exposed to the environmental strain of oil and gas production, and partly that Norway should reduce its total levels of oil and gas production because consumption of fossil fuels pollutes the environment. It is basically difficult to estimate how much of the future increases in oil and gas production will take place in vulnerable areas like the Skagerrak or the Barents Sea. For this reason, we have decided to analyse the effect of removing the LTP's precondition about future production of oil and gas from the categories 'field under consideration' and 'increased recovery and new prospects'. Without these categories production is gradually reduced relative to the reference path, so that 2020 production is around 70 per cent lower than in the reference path. This percentage is then held constant until 2030. In 2020, the production of oil and gas in the LTP is around 120 million tonnes of oil equivalents, possibly half of which is gas production. In alternative 1, production of oil and gas is reduced to around 40 million tonnes oil equivalents. In 2030, petroleum production in the LTP is a little more than 80 million tonnes oil equivalents (of which gas production accounts for around 80 per cent). Investment in the oil and gas industry has also been adjusted down by the percentage indicated in a report from the Ministry of Industry and Energy (1993). Investment linked to the production categories 'field under consideration' and 'calculation field' has been omitted in the impact calculation.

In this and the remaining MSG scenarios it is assumed that such unilateral Norwegian action will not have secondary effects on the Norwegian economy from a global market response in the form of price changes. Nor is any assumption made about changes in behaviour and strategic choices by recipients of Norwegian petroleum who could be thought to change CO_2 emissions outside Norway as a result of the policies indicated in these scenarios. This is quite clearly an unrealistic simplification because Norway is a significant player in the European energy market. The impact on energy consumption and production in Europe of a reduction in Norwegian gas exports depends on to what degree other gas producers replace the shortfall, and what the reactions to this sort of Norwegian policy would be in the importing countries.

There has been a certain focus placed on this area. Various theoretical studies of the impact have been carried out, yielding different results. Roland and Haugland (1995), in their report, find that a 10 per cent reduction of Norwegian gas exports will lead to global increases in CO_2 emissions in 2010, under the assumption of the reactions of certain market players, a given trend in energy demand and that there will be no international or regional climate agreement. Others, find that the total level of reduced Norwegian exports will have positive impact on the environment. The difference between the various results is large. Only a thorough empirical analysis of today's oil and gas export will provide us with the answer as to whether the activities on the Norwegian sector of the Continental Shelf are a positive or negative contribution to the global environment.

In alternative 1, GNP is reduced by 5 per cent compared to the LTP, a result of lower activity in the petroleum sector. This is a relatively small reduction, partly because oil production is relatively low in the reference path at the end of the projection period. The greatest reduction in GNP will take place just following the turn of the century, with a 9 per cent reduction in GNP in comparison with the level in the LTP. The balance in current prices deteriorates by around NOK 3500 million in 2030. In order to maintain the exogenously determined current account balance the net official transfer (interest payments and foreign aid) must change by the same amount and balance the deterioration in the trade balance. This would imply reduced transfers on, for example, foreign aid by NOK 3500 million which is a little more than 7 per cent. If volumes of exports and imports are examined, we see that imports are heavily reduced compared to exports. The reason for this is that relatively high levels of production factors in the petroleum sector are imported. Imports of crude oil increase, however, and by the end of the projection period Norway is a net importer of crude oil. At the same time, however, Norway is still a net exporter of refined oil products. As a result of reduced real income, the competitive edge in terms of costs is improved, and export is reduced little, thanks primarily to increases in exports of traditional goods, including agricultural and fish products, textiles, metals and other engineering products.

The model assumes full employment, so that reduced employment in one industry is balanced by increased employment in others. This can take place because the real income is reduced. Compared with the reference path, employment in such industries as wood-processing, production of manufactured raw materials, metals and engineering products increases. Employment also increases in the primary industries. In the cases of agriculture and fishing, however, exogenous production is still presumed in the model so that changes in employment patterns as a result of any changes in levels of production are not included in the calculations.

All consumer goods' prices fall compared to the reference path. Investment in the petroleum sector is also reduced. This shows a tendency towards an increase in consumption compared to the LTP. The reduction in income is, however, sufficiently strong for the consumption of all products to be reduced compared with the level in the LTP. Total private consumption is reduced by 9 per cent, much of which is made up of reductions in fuel consumption, car and housing conditions and trips abroad. Consumption of these products are especially heavily influenced by changes in income (high income elasticity).

Structural adjustment costs in alternative 1 – reduced oil and gas extraction

As MSG is a long-term model which, among other things, assumes exogenous employment and no unemployment, it is not possible to analyse the medium- and long-term processes of change which the Norwegian economy must undergo in the scenario which includes reduction in oil and gas extraction. In years to come, parts of the offshore-related industry will have to adapt to a generally lower level of activity and a change of focus in their products. In our alternative scenario, the process of change will be more difficult than it would have been, given the level of activity assumed in the LTP.

For this reason, we have used the complementary medium-term MODAG model to study the costs of intermediate structural change resulting from alternative 1. MODAG is more suitable for the study of problems connected to structural change. MODAG employs a simulated reference path, as well as an impact path which shows the effect of reduction in petroleum activity of the same size as in the MSG simulation.

In MODAG, tax revenues from the petroleum sector are to a great extent exogenous variables.[2] This means that reduction in petroleum extraction does not (strongly enough) lead to reduced revenues from oil taxation. If the model user does not pay explicit attention to this factor, the calculated effect on the public budgets will be wrong. Nor do the calculations take into account that the reduction in petroleum activity will necessarily require policy change.

In this case, we wish to primarily study trends in the labour market, paying attention to any unemployment during the period of structural change. It is important to be aware of the precondition that no measures to ease the process of change are to be implemented. This precondition will tend to make the calculations overestimate the impact on unemployment. On the other hand, a reduction in public tax revenues may lead to contractions, which tends to indicate that the calculations presented here underestimate the impact on unemployment.

Figure 18.3 shows that there are no dramatic differences between unemployment in the two scenarios. In the long term, the level of unemployment in the alternative converges with the level in the reference path. We also see that the number of unemployed resulting from the reduction in petroleum activities increases towards the turn of the century. The reason for this is the extraction, and especially the investment, profile for the petroleum sector in the alternative scenario compared with the reference path. Employment trends follow, to a very great degree, the trend in the difference between the reference path and the effect path. In the alternative scenario, the demand for investment objects from petroleum activities is presumed to become rapidly reduced compared with the reference path, and the absolute reduction in investment will be greatest in 2000. This will affect mainland Norway and lead to increased unemployment and reductions in the number of jobs available.

In 2000, employment will fall by about 30 000 as a result of reduced activity in the petroleum sector. In the long term, unemployment will

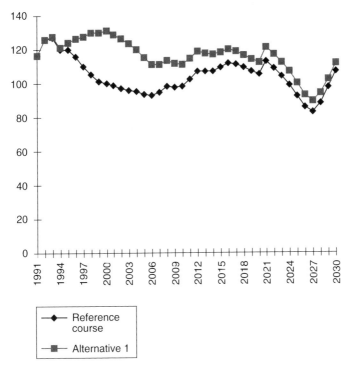

Figure 18.3 Number of unemployed in the reference path and alternative 1 (1 000 people)

fall, and as soon as 2010 the number of unemployed as a result of this lower activity will be more than halved compared with 2000. The reason is that, in the long term, petroleum activities in the reference path are reduced, so there will be smaller absolute differences between the reference path and the effect path. It may also be because Norwegian industry has managed, to a great extent, to adapt to the heavier reduction in petroleum activity in the effect-path alternative, and that employment has become less petroleum-related. All of the sectors show increased employment between 2000 and 2030. Certain sectors also have higher employment in the alternative scenario than in the reference path – for example, in production of consumer goods, of raw materials for the chemical industry and fishing and (in the long term) also in production of intermediate goods and investment products, in the construction and banking and insurance sectors.

Engineering products and ship- and rig-building will feel the effects of the reduction in the petroleum sector the most. In particular, sub-contracting and repair in the engineering sector will be reduced by around 60 per cent by 2000. Ship-building, especially oil platform-building, will be equally affected. The construction sector will be reduced by around 10 per cent. These reductions will lead to job losses. Jobs are also lost in domestic transport, other private services and trade. In all, employment levels fall by 3 per cent in 2000, 1.5 per cent in 2010 and 0.5 per cent in 2030, compared with the reference path.

Public sector budgets increase heavily. By 2000, the budget surplus of NOK 30 000 million has turned into a deficit of NOK 3000 million, 14 000 million of which is a result of a reduction in direct taxes on oil extraction. By 2030, the deficit in public sector budgets increases by NOK 142 000 million, or around 20 per cent. It is, however, important to remember the precondition that the income from state-run petroleum activities is unchanged.

Alternative 2: high CO_2 taxes

In alternative 2 a national CO_2 tax is levied on all emission sources. This CO_2 tax is levied in addition to all presently existing taxes which are present in the reference path (in 1993, the CO_2 tax on petrol was around NOK 320 per tonne CO_2 and on heating fuel it was NOK 140 per tonne CO_2). The CO_2 tax[3] is increased by NOK 300 per tonne CO_2 per annum in the period between 1990 and 2000, by NOK 200 between 2000 and 2010, and by NOK 100 per tonne CO_2 between 2010 and 2015, and after 2015 there will be no increase in the tax. Converted to 1989 prices, by 2015 the CO_2 tax will have been increased by NOK 7.10 per litre of oil. Because of inflation, the real value of the tax is NOK 4.6 per litre of oil in 2030. The main goal of this alternative is to achieve a significant reduction in Norwegian CO_2 emissions by levying equal tax levels on all types of emission. We have, however, ignored the effects these increases could have on oil and gas activities and on technological advances, which are both assumed to follow the same trends as in the reference path.

The CO_2 tax will have a special impact on those industries which are largely dependent on fossil fuels, or material which emits CO_2 in the production process. Price increases in energy-intensive products will be high in relation to less energy-intensive products. Consumers who use fossil fuels will also suffer higher costs. CO_2 taxes will, therefore, lead to changes in the composition of production and consumption.

The reduction in GNP in 2030 compared with the reference alternative is somewhat less than in alternative 1 (4 per cent). The increase in the CO_2 taxation has primarily consequences for the power-intensive industries. Production of ferro-alloys and aluminium are extremely energy-intensive. The metals sector will see its gross product reduced by 55 per cent in 2030 compared with the reference path – that is, the gross product is 3 per cent lower in 2030 than in 1989. These results are admittedly dramatic, although we must not forget that the simulation illustrates a situation which is 40 years ahead of the simulation baseline year. There are, however, a number of sectors which have experienced far larger relative change in the last 30 years.

The balance of trade in current prices deteriorates by NOK 700 million in 2030 compared with the LTP. If the current account balance is to be maintained, the balance of interest and grants must be improved by 700 million. As in alternative 1, aid is reduced, although the reduction is significantly lower than in alternative 1 (1.5 per cent). Measured in volume, total exports are reduced twice as much as imports. It is primarily imports from the energy-intensive industries (chemicals and metals) which are reduced in comparison with the reference path in this alternative scenario. Exports of other traditional products are also reduced. On the import side, imports of especially crude oil, refined petrol and heating oils are reduced. Employment in the chemicals and metals industries falls heavily. If employment is to be maintained, the level of real wages must be reduced by about 2 per cent and jobs increase in such industries as ship-building, production of intermediate goods and investment products, in the construction sector and agriculture. The gross real investments are reduced by NOK 7500 million, or 3.2 per cent. The total real capital is reduced by NOK 234 000 million (4.8 per cent).

Unlike alternative 1, consumer goods prices rise, with the exception of electricity. As a result of the CO_2 tax there are especially high rises in the price of heating oils and fuel. Households reduce their consumption of fuel by 65 per cent compared to the LTP. Compared with 1989, fuel consumption is reduced by 57 per cent in 2030. The consumption of heating oils is reduced by 44 per cent compared to the LTP in 2030, and by 30 per cent compared to the base year. Instead, households consume more electricity, which increases by 45 per cent compared with the base year and by 8 per cent compared to the LTP. However, the reduced activity levels in the energy-intensive industries allow for increased consumption of domestic electricity while maintaining the hydropower regime. The level of domestic consumption is 130 per cent higher in 2030 than in 1989 in the alternative path. Consumption is 2.3 per cent lower in 2030 compared with the LTP.

Alternative 3: combination of CO_2 tax and reduced oil and gas extraction

It would seem to be very unrealistic to assume that activities in the petroleum sector remain unaffected by the introduction of CO_2 taxes at such high levels as in alternative 2. For this reason, alternative 3 looks at a modified combination of alternatives 1 and 2. It can be assumed that petroleum activities develop as in alternative 1. The national CO_2 tax increases (over and above what is assumed in the basic alternative in the LTP) by NOK 100 per tonne CO_2 per annum in the period between 1990 and 2025. This is equivalent to a real tax increase of about NOK 3.30 per litre of oil in 2025 – in other words, a lower increase than in alternative 2. As we have noted, the question is primarily to examine the effects this will have on Norwegian CO_2 emissions and the Norwegian economy.

This combination of reduced petroleum activities and increased CO_2 taxation does improve upon the more unfortunate effects shown by a stand-alone increase in taxation (alternative 2), but it is still the power-intensive industry which suffers most from the measures. The mechanisms in the model are mainly as in the previous two scenarios, as the measures are the same. In this alternative, GNP is reduced by 8 per cent compared with the reference path in 2030. Consumption falls more; a reduction of a little over 10 per cent. The level of real wages falls by around 4 per cent in this alternative, compared to the reference path in the LTP. The balance of trade in current prices deteriorates by about NOK 4000 million, and the interest and grant balance improves by a similar amount (that is, NOK 4000 million less in aid).

Alternative 4: as alternative 3, but with exemption from CO_2 tax in power-intensive industries

Alternative 4 is the same as alternative 3 (increase in national CO_2 tax and reduction in petroleum activity), with the exception that the chemicals and metals industries are exempted from the CO_2 tax. The intention is to see the significance of these improved conditions for the Norwegian export industries on Norwegian emissions and the Norwegian economy. So, in alternative 4 we remove the tax increase for power-intensive industries.

The CO_2 tax exemption in the power intensive industries means that the reduction in GNP will be around 1 percentage point less than in alternative 3, while the reduction in private consumption is still around 10 per cent. Exports improve and imports are reduced somewhat less than in alternative 3. Measured in current prices, the balance of trade deteriorates by around NOK 4000 million, and aid is reduced by an equivalent amount (to maintain the current account).

Unlike alternatives 2 and 3, where the power-demanding industries reduce their production levels, in this alternative, production of metals and chemical products actually increases (at least, those exempted from the CO_2 tax). Activity levels in this scenario are between 30 and 50 per cent higher than in 1989. Trends in production of production materials and investment products do not vary significantly from the other scenarios. None of the demands on production is met in this alternative.

Alternative 5: as in alternative 3, but with reduced working hours

Alternative 5 is also the same as alternative 3, although it is also assumed that working hours are reduced by 10 per cent without wage compensation. In other words, we maintain measures aimed at the petroleum sector and increase the CO_2 taxation, including for the power-intensive industry. This alternative is meant to illustrate what the significance would be if future workers placed so high a priority on leisure that they voluntarily refrained from working as much as they do today. Reduced working hours, taken in addition to the measures in alternative 3, lead to a reduction in GNP nearly twice the size of that in alternative 3 (20 per cent). Consumption is reduced even more (22 per cent). Compared with 1989, however, GNP increases by 50 per cent and consumption by 82 per cent. Foreign trade does not do so well; exports are reduced by 17 per cent, measured in fixed prices, while imports are 'only' reduced by 13 per cent. Primarily the export of metals and chemicals is affected, although the export of other traditional products is also reduced. This is the scenario where aid is reduced most heavily in order to retain the current account balance (by 12 per cent, or NOK 6000 million according to the reference path). Real wage levels increase in this alternative by a little more than 2 per cent. CO_2 emissions are reduced by 40 per cent compared with the reference path.

18.3   Key findings from the alternative scenarios

In this section we will sum up and compare the main results from various future scenarios which emphasise the goals of the 'Environmental Parliament'. The calculations are presented in greater detail in Alfsen, Larsen and Vennemo (1995). This paper also discusses the significance of the petroleum sector on the Norwegian economy in more detail. Which can be useful for an understanding of how the models react to change in the petroleum activity.

The various scenarios are balanced against each other, both for sustainability and other effects on petroleum activity. When the alternative simulation paths are compared with the reference path (LTP) it is important to be aware that the MSG model assumes full employment and the same trends in the current account balance[4] as in the LTP. For this reason, the results of the simulation show what economic growth is necessary, and how this growth must take place if Norway is to achieve full employment and satisfactory trends in the current account balance in the future. Economic growth cannot be taken for granted, as a result of the policy measures which are introduced. On the other hand, long-term increase in macro-economic problems cannot be taken for granted as a result of the measures. In all of the simulations it is given that the surrounding world remains unchanged compared to the trends sketched out in the reference path. In other words, we are just looking at the effect of unilateral Norwegian measures, and ignoring any foreign reactions to our measures. This means, among other things, that world market prices of crude oil and gas do not change between the alternatives.

Table 18.3 presents some of the main results from the alternative future scenarios. The table shows percentage changes in some of the macro variables, as well as emissions compared to the LTP in 2030. Economic growth is somewhat lower in the alternative-effect paths than in the reference path, which shows a tendency towards less polluting emissions. In addition, the emissions are influenced by changes in industrial relationships in the alternative scenarios. The changes in levels of emissions vary heavily between the components. This reflects,

Table 18.3   Changes from the reference path, in 2030 (per cent)

Alternative	1	2	3	4	5
GNP	−5	−4	−8	−7	−18
Private consumption	−9	−2	−10	−10	−22
Exports	−1	−9	−7	−4	−17
Imports	−3	−5	−6	−5	−13
CO_2	−15	−30	−35	−26	−40
NMVOC	−30	−15	−40	−40	−47
NO_x	−10	−15	−20	−17	−26
CO_2	+2	−40	−32	−12	−35

to a very great extent, the petroleum sector's share of emissions of various components. Emission of non-methane volatile organic components (NMVOC) are affected most heavily by reductions in petroleum extraction, as the NMVOC emissions from petroleum extraction make up a total of 40 per cent of total emissions. Alternative 2, which presumes increased CO_2 taxation, will in this way achieve the lowest reduction in NMVOC emissions. The increased emissions of SO_2 in alternative 1 is primarily a result of increased production in the metals sector, which is responsible for around 30 per cent of total emissions. Emissions of nitrous oxide and methane originate mainly from agriculture, and are affected little compared to the LTP.

The petroleum solution (alternative 1) primarily affects NMVOC emissions. CO_2 emissions are somewhat reduced, although SO_2 emissions increase. The CO_2 tax (alternative 2) works well with respect to CO_2 and SO_2 emissions but, when taken alone, has negative results for power-intensive industry. A combination of CO_2 tax and reduced petroleum activity (alternative 3) is effective in reducing emissions of NMVOC, CO_2 and NO_x, but GNP and private consumption become fairly heavily reduced. Exemption for the power-intensive industry (alternative 4) increases CO_2 emissions compared to the alternative with no exemptions (alternative 3), while SO_2 emissions increase heavily. NMVOC emissions are not affected by exemption of the power-intensive industry from an increase in CO_2 tax. Reducing working hours by 10 per cent (alternative 5) leads to a reduction in all emissions by between 3 and 7 per cent compared with alternative 3, and this is the alternative which gives the greatest reduction in emissions.

In Chapter 23 we will return to a more detailed discussion of the results in the light of the 'Environmental Parliament' goals.

18.4   The impact of CO_2 taxation (alternative 2)

Table 18.4 shows some of the principal results for alternative 2. The balance of trade in current prices falls by NOK 700 million in 2030 compared with the LTP. If the balance of trade is to be maintained, the interest and grant balance must be improved by 700 million. As in alternative 1, this means that, for example, aid must be reduced. Total exports in fixed prices are reduced by NOK 50 000 million (9 per cent) and imports in fixed prices are reduced by NOK 26 000 million (4.5 per cent). Exports from the power-intensive industries (chemicals and metals) are primarily reduced in this alternative scenario when compared with the reference path. Among imports, it is especially

Table 18.4   Deviation in some of the main figures from their equivalents in the 'basic alternative' (LTP), 2030

	Level change (NOK 1989 1 000 million)	Percentage change from LTP, 2030
GNP	−47	−4.2
GNP growth	−0.10 (% per annum)	
Gross product, petroleum extraction	+1	
VAT, etc.	−5	−3.3
Private consumption	−17	−2.3
Consumer Price Index		0.0
Balance of trade, current prices	−1	
Exports	−50	−9.0
Imports	−26	−4.4
Real gross investment	−8	−3.2
Real capital stock	−234	−4.8

crude oil, petrol and heating fuel imports which are reduced. Gross real investment falls by NOK 7500 million, which is 3.2 per cent. The total real capital stock is reduced by NOK 234 000 million (4.8 per cent).

Employment figures fall in the chemical industry and metals. If employment levels are to be maintained, the real wage level must be reduced by around 2 per cent, and jobs created in such industries as ship-building, construction, agriculture and fishing. Prices for most consumer goods increase, apart from electricity. There is a particularly heavy price increase in heating fuel and fuel resulting from the CO_2 tax. Households reduce their consumption of fuel by around 65 per cent in 2030 compared with the LTP (see Table 18.5). Compared with 1989, consumption is reduced by 57 per cent in 2030 in the alternative curve. The number of kilometres driven is not, however, reduced as much as a result of considerable technical advances in this sector; personal mobility is thus much less affected. Consumption of heating oil is reduced by 44 per cent compared with the LTP in 2030, and by 30 per cent compared with the base year. Households begin to use more electricity instead. Consumption of electricity increases by 45 per cent, compared with the base year and by 8 per cent compared with the LTP. Reduced activity levels in the power-intensive industries make it possible to increase domestic consumption of electricity within the extents of the hydropower regime. The level of private consumption is 130 per cent higher in 2030 than in 1989 in the alternative path. Consumption, compared with the LTP, will be 2.3 per cent lower in 2030.

Table 18.5 Percentage change in private consumption, 2030 (LTP), 2030

Consumer goods	Percentage change compared with 1989	Percentage change from the reference path, 2030
Electricity	46	+8
Oil for fixed installations	–28	–44
Petrol	–57	–65
Car purchase	235	–1
Public transport	132	–4
Food	65	–1
Beverages and tobacco	121	–1
Other products	149	–1
Clothing and footwear	160	–0
Furniture	201	–1
Housing	100	–0
Other services	205	–0
Foreign tourism	269	–0

Distribution effects according to household type

Taxation of CO_2 emissions changes both consumer income and prices. Different households have different consumption patterns, and the impact upon them is therefore different when relative prices change. Increased CO_2 taxation on fuel and heating oil affects different households in different ways, depending on the share of their budget devoted to fossil fuels. Generally speaking, the share of the budget set apart for fuel increases in line with income, meaning that households with the highest incomes will be hardest affected by CO_2 taxation. On the other hand, the share of the budget spent on heating oil decreases in line with income, so that heating costs more, relatively speaking, for low income groups.

Aasness *et al.* (1995) show that the price impact on a wealthy household is less than for a poorer household, because the share of their budgets devoted to oil products is greater in the poorer household. Totally speaking, the poorer households are more heavily affected by the CO_2 tax than the wealthier. They also find that childless households also suffer a greater reduction in welfare than families with children.

Estimates of the benefit of emission reductions

Statistics Norway has created a simple accounting routine which, based on changes in air quality, estimates simple economic gains or losses connected with these changes. The basic data which is used to calculate the changes in emissions to air and road traffic in monetary terms, which are connected to changes in damage to nature, materials and people, as well as noise, queue formation, traffic accidents and road damage, have been collected from many sources, both Norwegian and international. In the case of health effects, the Norwegian Pollution Control Authority, aided by medical experts, has estimated the size of the costs involved in the exposure of one person to pollution concentrations above medically set threshold values. The distribution model is used to calculate how many people are exposed to concentrations of pollution over the threshold value in an emission increase. This creates a link between changes in emissions and changes in the incurred health costs. For the time being, the basis for these damage estimations is admittedly uncertain. The estimates presented here do, however, provide an indication of the size of savings which can be made when emissions and traffic are reduced. (For a more detailed discussion, see the Project for a Sustainable Economy, *Report* 15.)

Table 18.6 estimates the benefit from reductions in emissions and traffic in alternative 2. If we allow for the uncertainty in the underlying data, we find that the reductions in environmental costs which follow the introduction of a high CO_2 tax reaches a total of between NOK 1000 million and 13 000 million. In comparison, GNP is reduced by NOK 47 000 million.

Table 18.6   Estimated benefits from emission reductions in alternative 2 compared to the baseline in year 2030 (NOK = million, 1989 prices)

Type of damage	Benefit (cost)
Acidification of water	10
Acidification of forests	26
Corrosion costs	90
Health injury from NO_x	2 263
Health injury from SO_2	101
Health injury from CO	3
Health injury from particles	207
Traffic accidents	567
Access	608
Road damage	759
Traffic noise	282
Total	**4 916**

Table 18.7   Benefit of reduction in emissions and traffic as well GNP reduction compared with the LTP for 2020 (NOK 1 000 million, 1989 prices, and percentage changes)

	Alternative				
	1	2	3	4	5
Benefit, range	1–9	1–13	2–24	2–24	2–33
Benefit, point estimate	3.5	4.9	10.1	9.8	13.8
GNP (%)	–56 (–5)	–47 (–4)	–90 (–8)	–76 (–7)	–205 (–18)
Ratio between benefit and GNP (%)	6.25	10.4	11.2	12.9	6.73
GNP corrected for environmental harm (%)	–4.5	–3.5	–7.0	–6.0	–17.0

The benefit of reductions in emissions and in traffic

Damage calculations (Table 18.6) are uncertain both in regard to the benefits connected with changes in emissions and in traffic, and as far as favouring petrol and diesel fuel consumption for road transport rather than other forms (for example, ships and other vessels). Table 18.7 shows the uncertainty range for reduction in environmental damage in addition to GNP reduction in the alternative scenarios. To make it easier to compare the alternative scenarios for benefit of reductions in emissions and road traffic, Table 18.7 also shows a point estimate which is located between the upper and lower range limits.

A separate analysis of MODAG has not been made which illustrates the costs of structural adjustment involved in a CO_2 tax of the size indicated in alternative 2, as this would require significant modelling techniques. Analyses have, however, previously been made (see Moum, 1992) which can indicate something of the necessary changes involved in a national CO_2 tax. The KLØKT calculations indicate that a national stabilisation of CO_2 emissions (thanks to the CO_2 tax which forms the basis for reduced employers' contributions) can be carried out without any great macro-economic costs in the form of reduced consumption, production or employment. The trade balance would, however, be adversely affected to a certain extent. As in the MSG calculations, the MODAG calculation indicates that some significant changes in the structure of the industrial base are necessary, from the industries with heavy CO_2 emissions per unit produced to the industries which emit relatively little CO_2. The measures used result, for example, in an appreciable reduction in the power-intensive industries, while production in the labour intensive industries increases.

If the introduction of a CO_2 tax takes place quickly, the adjustment costs involved may be great. A slower introduction of the measure over a long period would reduce the adjustment costs, if it is credible for the measure to be introduced completely.

If the lower limit for environmental costs is chosen, the value of emission and traffic reductions (reduced congestion and accidents) will be only NOK 1000 million in alternatives 1 and 2, NOK 2000 million in alternatives 3, 4, and 5. The highest estimate gives a value of emission reductions of between NOK 9000 million in alternative 1 and NOK 33 000 million in alternative 5. For this reason, it is decisive for the results which values linked to the emission and traffic reductions one believes in. The GNP reduction which is corrected for the benefit of a better environment will be between 0.5 and 1 percentage point less (based on the point estimates) than the non-corrected reduction. Alternative 5 provides the greatest reduction in emissions, and therefore has the greatest impact on environmental improvement. The environmental benefits can, however, far from compensate for the considerable GNP reductions in alternative 5. Furthermore, this fifth scenario results in an accelerated deterioration of the budget balance, thus transferring a major part of the costs of present generation spending to future generations.

National wealth and real income

Future consumption options and welfare will depend on how the development of the total national wealth – that is, how the human capital (education, state of health), production and consumption capital, debts or claims on foreign economies and the reserves of natural resources – is managed. Such wealth variables as real capital, petroleum wealth and claims against foreign economies are the easiest

Table 18.8   Estimated wealth for 2030 in the reference path and the alternative scenarios (1 000 NOK per capita)

Wealth	1989	LTP	Alternative				
			1	2	3	4	5
Real capital holding	545	1 021	961	971	924	940	845
Net foreign claims	31	113	115	113	115	114	112
Petroleum wealth	200	333	58	33	58	58	58
Total	776	1 167	1 134	1 117	1 097	1 112	1 015

to place a value on. Table 18.8 shows these three components in the Norwegian national wealth in the LTP and the alternative scenarios.[5]

In the LTP, the total of the three components in 2030 is at least 50 per cent greater *per capita* than in 1989. The trend in man-made capital reserves dominates wealth development in Norway. In the alternative scenarios, the wealth is somewhat lower than in the reference path, although it still shows growth of between 31 per cent and 46 per cent in the period from 1989 to 2030. All the alternative scenarios, with the exception of alternative 2, presuppose a reduction in petroleum extraction. This means that, by the end of the forecast period, Norway has an even greater remaining petroleum wealth compared with the LTP. Alternative 5 gives the lowest national wealth measured in real and financial capital and the petroleum wealth. There is, however, no very great difference between the alternatives.

The main source of increased material wealth in the forecasts comes, however, from human resources and technological advance. In the LTP, the growth in return on the petroleum wealth and man-made capital can explain only around 30 per cent of the growth in disposable real income *per capita* from 1989 to 2030. Figure 18.4 shows the estimated national wealth in 1991. A highly qualified workforce is Norway's most important economic resource.

Figure 18.5 shows the disposable real income for Norway (GNP *plus* surplus on the interest and grant balance *less* the wear and tear on capital) in the reference path (the LTP) and the various scenarios. Figure 18.5 shows that Norwegian real incomes are not changed to any significant extent in the alternative scenarios. The reduction in GNP shows a tendency towards reduced disposable real income for Norway in the

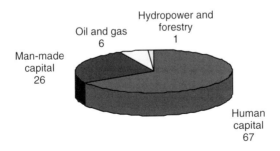

Figure 18.4   Estimated national wealth in Norway, 1991, according to source (per cent)
Source: Alfsen, Larsen and Vennemo (1995).

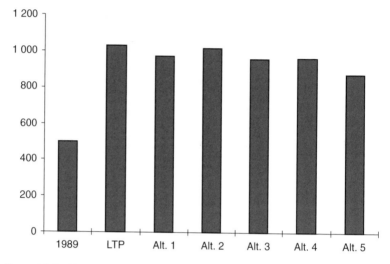

Figure 18.5   Disposable real income for Norway in the various scenarios, 2030 (NOK 1 000 million)
Source: Alfsen, Larsen and Vennemo (1995).

alternative scenarios, while improved interest and grant balance (at the expense of aid) and *less* wear and tear and capital shows a tendency towards an increase in real income. All in all, the reduction in disposable real income in Norway, compared with the reference path in 2030, varies between 1 per cent in alternative 1 and 15 per cent in alternative 5. There is, however, as in the LTP, relatively strong growth in disposable real income from 1989 to 2030 in all of the alternative scenarios.

Government budgets

Balance in government budgets in the reference path are roughly sustained throughout the entire simulation period. There is a slight surplus in 2010, which falls to a slight deficit in 2030. The deficit is 1 per cent of GNP in 2030. The impact on public budgets in the alternative paths is difficult to comment realistically upon without binding trends in variables which are politically determined. Our comments about public budgets here must not, therefore, be regarded as anything else than a description of the first-order impacts in budgets when policies are assumed to remain unchanged (that is, in the LTP).

Generally speaking, the trends towards lower growth, with its lower taxable base, mean lower surpluses in the alternative scenarios. Alternatives 2 and 5, in particular, deviate from the public sector budgetary trends in the LTP. In alternative 2, incomes rise as a result of the relatively strong increase in CO_2 tax compared with costs until 2010. The real value of the tax falls after 2010 and the taxable base becomes less, and by 2030 the deficit as a percentage of GNP is as predicted in the LTP. There is also a comprehensive redistribution between local and national authorities in alternative 2. The state receives major revenues from CO_2 tax, while local authorities are affected by the generally lower taxable base. Major deficits in local authority budgets, while state budgets are well in the black, can be avoided only by heavy state funding of local authorities when CO_2 tax is increased. Alternative 5 is also noticeably negative as regards public sector budgets, as a result of the predicted reduction in working hours which, hand in hand with reduced growth in GNP, means that the state net debt increases heavily, as do interest charges. As a result of the reduction in working hours and the loans necessary for the funding of public programmes, the public sector faces a severe financial problem in this alternative. The public sector deficit is 11 per cent of GNP in 2030. There is also an unfortunate effect on national debt in alternative 1, although not as severe as in alternative 5. The deficit, as a percentage of GNP in alternative 1, increases as the turn of the century approaches, and then reduces. Despite the increase in interest charges in alternative 1, income still exceeds expenses in 2030, mainly as a result of lower transfers to the domestic economy and foreign economies. Taxable revenues from petroleum extraction are reduced by 70 per cent in 2030 (around NOK 13 000 million) compared with the LTP in all alternatives, apart from alternative 2 (where there is a slight increase).

Part V

Additional Future Perspectives with Complementary Analyses

19
The Presumed Impact of CO_2 Taxation on Various Sectors

19.1 Partial sector impacts

Chapter 12 shows, on the basis of both economic theory and historical fact, that it is possible to create a double dividend through various environmental measures directed at Norwegian industry. In many cases there is significant unexploited potential for replacement of polluting materials/processes by more environmentally sound alternatives, while at the same time processing industrial waste, which is after all a traditional environmental problem, into a profitable product (see Chapter 15). Environmental taxation can be an economically efficient way of accomplishing this. At the same time it can be used to replace other, more distortionary (and therefore economically less efficient) taxation. In many cases, Norway can introduce these unilaterally, because they can prove to be a long-term strength to the competitive edge of Norwegian industry.

When industries enjoy various forms of market protection and support – for example, long-term power contracts at prices which are far lower than marginal power supply costs – it is reasonable to assume that this sort of environmental taxation could release unrealised resource saving. In the case of CO_2 emissions, the possibility for such substitution is, however, more limited. CO_2 taxation could have exactly the opposite effect on those parts of the Norwegian economy which have already adapted efficiently to the competitive market.

Several studies have examined the impacts of the introduction of CO_2 taxation on Norwegian energy-intensive industries (including Bye, *et al.*, 1989; Mathiesen, 1991; Mathiesen, 1992). If we are going to discuss these effects in relation to the goals of the 'Environmental Parliament', then it is important to define the following:

(a)   The difference between a CO_2 tax on:
    (1)   Solely fossil fuels.
    (2)   All CO_2 emissions (including the process industries).
(b)   Do the proposals apply to:
    (1)   A unilateral Norwegian CO_2 tax (that is, no other country also introduces one)?
    (2)   A CO_2 tax in all OECD countries?
    (3)   A global CO_2 tax?

The impact of a CO_2 tax, both nationally and internationally, may vary widely, depending on which of the above conditions is assumed to apply. Generally speaking, the studies have shown that a unilateral Norwegian CO_2 tax will have a negative effect on profitability (often strongly negative) for those industries which consume great amounts of fossil fuels and/or are dependent upon them in industrial processes, even in the reasonably long term. Many of the raw material processing industries are very sensitive to changes in the economy, and choice of which to analyse is important. We can, however, establish fairly generally, on the basis of these studies, that a unilateral heavy Norwegian CO_2 tax will lead to reduced CO_2 emissions in Norway, just as we have shown in the MSG model long-term scenarios in this study. It is not necessarily a fact that the reduction in emissions results from reorganisation of production with the aim of providing lower fuel consumption and less processing. Some of these industries may be wound down or moved, perhaps transferred to manufacture in countries which have not introduced such equivalent CO_2 taxation. In other words, we cannot exclude the possibility of the global CO_2 impact of unilateral Norwegian taxation being, in the long term, an increase in the number of less efficient producers abroad, even if Norway, taken in isolation, could meet its national CO_2 targets through such a measure. This conclusion could apply for such industries as:

- fish oil and meal
- cement
- carbides
- ferro-alloys
- primary aluminium.

In the case of the wood-processing and ferro-alloy industries, biofuel and wood chips are alternatives to fossil fuels, although the price of these alternatives is initially much higher, so that profitability at the time of substitution will be drastically weakened.

A CO_2 tax on fossil fuels throughout the OECD countries will lead to major simultaneous reorganisation problems for the power-intensive industries in these countries, although it would be much easier for Norwegian industry, at least in comparison with a unilateral Norwegian imposition of the tax. In the long term, even in the case of a regional introduction of the tax, Norwegian industry would have to be prepared for loss of market share to competitors outside the OECD. As a result of the probability of movement outside the OECD area and/or increased investment in the Third World, the probable outcome is that global CO_2 emissions would increase.

A global CO_2 tax would be the best solution for both the climate and the environment, while being also the politically most difficult solution. It would mean that transfer to an 'emission paradise' would cease to be an alternative for these industries, and those companies which had invested most heavily in energy-saving technology and processes based on substitutes for fossil fuels, would achieve a competitive edge. In this case, Norwegian industry would mainly emerge as the strongest competitor when compared with other OECD countries and the rest of the world.

19.2 Activity and employment impacts of reduced CO_2 emissions[1]

In the MSG and MODAG models, Norwegian production is divided into a metals sector and a chemicals sector. This is quite clearly highly aggregated considering the division of industries which are specially important with respect to the extent of the emissions that we are especially concerned about monitoring. If we are to say more about the links between reduced CO_2 and employment, apart from the conclusions we can draw from the alternative scenarios in the MSG and MODAG models and the above theoretical considerations, we must use a complementary model where these two industries are disaggregated into five different sectors. On the other hand, other manufacturing and service sectors, which are assumed to have lesser significance for CO_2 emissions, are compressed more than in MSG and MODAG.

The SNF model adapted for this exercise is a general equilibrium model in the same way as MSG, but has a much shorter time horizon (around 10 years), which is closer to the MODAG model. The model shares the same assumptions about frictionless adaptation in the labour market as MSG – that is, that when new equilibrium is established, unemployment does not exist because the wage formation is such that it cannot exist. One objection raised against this choice of model is that it is a very unrealistic description of wage formation in Norway. 10 years is also a somewhat short period of transformation in which to achieve full employment after a reform of the economic structure which places a far greater emphasis on environmental taxation than we do today.

Another important difference between this model and MSG/MODAG is that while the latter are based on historically estimated links in Norway, the SNF model is based on assumed values. This means that it does not provide a basis for empirical assertions in the same way as the MSG/MODAG models. The argument for using such a complementary model for further illustration of this important question, even so, is that analyses have shown that choice of model structure and sector dis-aggregation can be vital for the strength of an analysis' assertions when we are specially concerned with a more detailed specification of trans-port and the processing industry. The model describes production within about 30 sectors and the input–output flows between them. The sectors differ in their treatment of factor use and emission sizes which, in the model, are affected to different degrees by the goal of reducing total emissions. The costs and budget items used as a basis for calcula-tions are taken from the 1991 national accounts, while the figures from the MODAG calculations at the Ministry of Finance have been used to update the model from 1991 to 2000.[2]

International trade

The open Norwegian economy, and the fact that production of many of our largest export items causes high CO_2 emissions, makes model-ling international trade – and especially export – very important. We assume that export of these products faces falling demand. Price flexi-bility has been roughly fixed on the basis of SNF's studies of several of the markets (Mathiesen, 1988a, 1988b, 1990b), and reported estimates from various sources, including Statistics Norway and economic theory. Generally, SNF's price flexibility seems to be somewhat higher (numerically) than those appearing in Statistics Norway's analyses.

The estimates have been carried out for a unilateral Norwegian reduction in CO_2 emissions, but give us a full insight into how the export industries would be affected by an international agreement on regulation of emissions of greenhouse gases. As of today, it is fairly unclear how such an agreement could be drawn up. All we can do here is to point out that the models allow a certain (although far from com-plete) price transfer in the global market. Manne and Mathiesen (1994) analysed the consequences for the global aluminium industry of a carbon tax within the OECD area, and concluded that between 30 and 40 per cent of the tax was compensated for in the price of aluminium. The rest of the tax resulted in reduced prices for production factors. The reason was that all of the expansion in the industry took place outside the OECD area, and that these factories, using the most modern technology, provided competition which pressed down prices in the market. This movement of production outside the OECD has taken place over the last 20–30 years, and a carbon tax within the OECD would only increase its pace.

Imports have been treated somewhat differently, and in significantly more detail, as the cost functions of each of the consuming model sectors contains choices between Norwegian and imported products. The most significant weakness in our analyses is that we can study only unilateral Norwegian measures. This does not allow us to take into consideration the fact that production of imported products also, to a greater or lesser extent, involves emission of greenhouse gases and that these products will also increase in price in the case of an international agreement, allowing Norwegian products to face softer competition.

The final point about the foreign economy is that we stipulate an expected surplus in the operating accounts. This includes a forecast deficit in the interest and grant balance, and a larger export surplus. We have also analysed the effect of the surplus on equilibrium.

A political goal for the total emissions in the economy makes the sectors mutually dependent. If the goal is to be achieved, some of them must reduce their emissions, and the more *one* sector reduces its emissions, the easier it becomes to achieve the goal, lessening the pressure on the others.

Mathiesen (1991) has also studied the position of a Norwegian CO_2 reduction as part of an international agreement. The assumptions were in line with the estimates of the environmental tax committee, that prices of many Norwegian export products would rise to a certain extent to reflect severe cost transfer of national CO_2 taxes in the international market, while the oil price was predicted to fall by 20 per cent as a result of, among other factors, reduced international demand. The result of these calculations was that while a unilateral 15 per cent Norwegian reduction would cost 0.5 per cent of private consumption, an international agreement would cost 5 per cent! The difference results from lost oil revenues and illustrates the significance of Norwegian oil exports. We maintain the assumption of unilateral Norwegian CO_2 reduction, although the analysis shows us the consequences of an international agreement in which Norway, for example, was obliged to carry out equivalent limitations. (Compare the previous discussion on modelling export demand.) Instead of modelling the emissions as such, the model addresses the right to emit, as we assume that all of the players (sectors) have to acquire such a right. The emission right is therefore regarded as a factor in production and consumption which is analogous to other factors. In the model, the sectors are mutually dependent through the market for the emission rights which will be created. It is also assumed that the state issues and sells these rights, as a method of raising revenue. A central assumption for the analysis is that all sectors will pay the same tax per emitted unit, and the state will earn revenue from this. We shall, however, also develop scenarios which grant complete or partial exemption for certain industries, and where these exemptions can be linked to other exemptions granted by the Norwegian Pollution Control Authority.

Sector impacts on industry and employment

Some of the emission sources must be regulated if the goals of reduced CO_2 emissions are to be reached. Efficient regulation is based on the sources which have large emissions compared with their value added. Figure 19.1 demonstrates the appreciable difference between the sectors in this case, showing emissions of CO_2 per NOK gross production value. The emissions are shown in three parts: CO_2 emissions

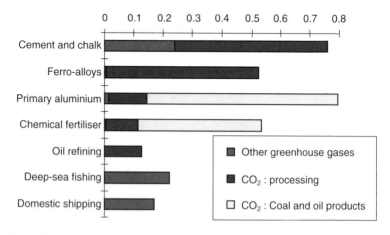

Figure 19.1   Emissions of CO_2 and other greenhouse gases (kg CO_2 equivalents), per NOK gross production value, selected sectors

from, respectively, processing and combustion of fossil fuels, and emission of other greenhouse gases converted into CO$_2$ equivalents. Industrial processes pollute heavily, even compared with service industries in the transport sector. In pure CO$_2$ analyses, other emissions of greenhouse gases would not be included. Figure 19.1 indicates that the distinction between regulation of only CO$_2$ emission, and regulation of all greenhouse gases based on CO$_2$ equivalents, may be significant for the production of primary aluminium and chemical fertiliser.

The nature of the model means that we do not place emphasis on individual figures, taking into account only qualitative trends in the solutions.

Various ways of using the revenues gathered from a CO$_2$ tax will impact in different ways on the economy. We will examine two alternatives a little more closely:

(a)   Transfer of all of the revenues from a CO$_2$ tax as a lump sum grant to households (lump sum alternative).

(b)   Use of the revenues from a CO$_2$ tax in their entirety to reduce employers' contributions (taxation alternative).

Figure 19.2 shows a percentage reduction in the level of activity per sector for those sectors which are heavily affected by a 20 per cent CO$_2$ reduction in 2000, compared with the reference path. There are insignificant differences in the results from the lump-sum alternative and the taxation alternative in this respect, because the sectors use little labour, so that lower payroll costs do not to any great extent compensate for the increase in CO$_2$ taxation. Refining and production of ferro-alloys is quite clearly the greatest loser: this industry is almost halved. Then follow cement, chemical fertiliser and primary aluminium which all suffer a 15–25 per cent reduction. Consumption of heating oil and petrol are also reduced equivalently. By far the least reduction (5–10 per cent) can be observed in the sectors for production of raw materials for the chemical industry, non-ferrous products and other crude oils, in deep-sea fishing (note that the model distinguishes between ocean fishing and aquaculture) and in the transport sectors domestic shipping, car rental and air traffic. Reductions in other sectors are insignificant, some even return a small increase.

The relative difference between the scenarios for these sectors for trial use is greater (see Figure 19.3). Reduced employers' contributions stimulate to an increased level of activity, and while most sectors meet recession in (a) because the CO$_2$ regulation leads only to increased

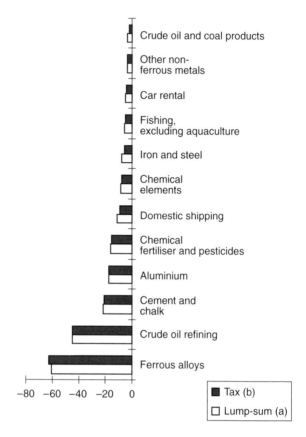

Figure 19.2   Percentage reductions in activity levels for the hardest-hit sectors in the case of equal taxation for a 20 per cent reduction in CO_2 emissions

costs, several see a small increase in (b) – for example, other industry increases by 0.7 per cent in (b) while activity levels are reduced by 1.2 per cent in (a). This stimulant creates far more jobs than are lost in the processing industry. Total employment levels in (b) are 2 per cent higher than in (a), and we can see from Figure 19.3 that most sectors enjoy around 2 per cent higher activity in (b) than in (a). The electricity sector is a special case. Production (high-tension power) is reduced as a result of reduced activity in the power-intensive sector, while turnover in mains power (low-tension power) rises somewhat. (Export also increases.)

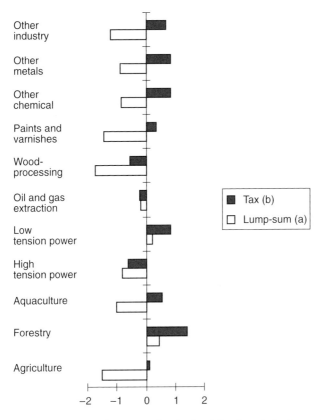

Figure 19.3   Percentage reduction in level of activity for relatively unaffected sectors in the case of the same tax, for 20 per cent reduction of CO$_2$ emissions

The occasionally dramatic reductions in the level of activity for the emission-intensive sectors has, of course, consequences for their employment levels. One precondition of these analyses is that those employees who lose their jobs find work in other sectors. While it is easy to spotlight those (few) sectors which are clear losers if the same level of CO$_2$ taxation is imposed on all, there are no equivalent clear winners. Labour is absorbed in the model's large 'omnibus' sectors, 'other industry' and 'other private services'. Expansion takes place, naturally enough, in the sectors which are least affected directly by the environmental taxation, or indirectly through their production factors – that is, those sectors which contribute least to emissions, but are stimulated by reduced employers' contributions. The model places the

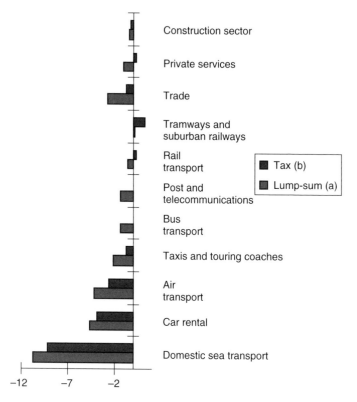

Figure 19.4   Percentage reduction in level of activity for service sectors, including transport, in the case of the same level of taxation, for 20 per cent reduction of CO_2 emissions

greatest load on those sectors which either pollute greatly or use heavily taxed products as production factors. It is exactly this change from emission-intensive to labour-intensive activity which is sought.

The percentage changes in employment in those sectors which are most affected is formidable, being halved in the case of refineries and production of ferro-alloys. It is, however, important to note that these industries form a very small proportion of total employment in Norway, in fact only 0.3 per cent. The total number of jobs in production of cement, chemical fertiliser and primary aluminium, which is responsible for 0.6 per cent of all jobs, falls by 10–15 per cent. Over a 10-year period this is no more than is commonly referred to as 'solving structural adjustment through natural attrition in the workforce'.

Table 19.1 shows these industries' share of jobs in the model's reference solution in 2000 – that is, with no CO_2 taxation and the job reductions in the case of a 20 per cent reduction in total CO_2 emissions. In total, employment increases by 1.1 per cent (in alternative (b) compared with the reference path and at least 2 per cent compared with alternative (a)), where employment figures fall compared to the reference path, and in the great mass of industries which are responsible for almost 96 per cent of all of Norway's jobs, it increases by 1.3 per cent on average. The problem is, of course, that small and isolated centres suffer a certain amount of the job losses.

Changes in the model's prices signal the extent to which a product contributes to CO_2 emissions. The SNF model is a little too simple in this respect. In the same way as in MSG/MODAG, the possibility of substitution between different products as factors in the individual sector is ignored. In this way, the opportunity for a customer of a given production factor to substitute others is underestimated. This is probably one of the greatest weaknesses in a number of general equilibrium models used for CO_2 analyses. In the case of cement, in particular, we have modified the expression so that cement is in a substitution relationship to lumber on the demand side – for example, the price of cement increases by 80 per cent and the price of ferro-alloys by 30 per cent. The reason that production of cement falls considerably less than production of ferro-alloys is that while cement is sold in a virtually protected domestic market, where the increase in costs can be largely transferred to the price

Table 19.1   Impact on employment of a 20 per cent reduction in total CO_2 emissions

Industries	Employment share (%)	Changes in employment levels (%)
1. Refining and ferro-alloys	0.3	–50
2. Chemical fertiliser, primary aluminium and cement	0.6	–12
3. Chemical raw materials, iron foundries and domestic shipping	2.1	–4
4. Deep-sea fishing, other crude oils, non-ferrous metals, car rental and air travel	1.3	1.1
5. All other industry and public sector	95.7	1.3
Total	**100.0**	**1.1**

of the product, ferro-alloys are sold in a competitive international market where the unilateral Norwegian costs can be transferred to the price only to a limited extent. The increase in costs for ferro-alloys will, therefore, lead to an appreciable reduction in exports. One result of this will be to move production and jobs abroad, probably leading to greater emissions from foreign production. This will be detrimental to the climate, at the same time as Norway loses jobs. Similar conditions obtain in the case of refining. Refined products (petrol and heating fuel) are presumed bought and sold on the global market. The domestic increase in costs will therefore result in heavily reduced volumes.

In the model, export products are divided into relatively detailed sectors and the demand for them is elastic. There is greater uncertainty around the quantification for the price sensitivity. Greater price elasticity is primarily an effect only for ferro-alloys which experience even greater reductions. Why does little happen in the other export product sectors? The logic of the model is that the economy consists of a number of sectors which convert various factors, including emission rights, into money and it sorts the sectors according to their profitability as wealth creators: production of ferro-alloys is the least profitable! The emissions of the sector are too large in comparison to its value added. It is also worth noting that the industry has returned very weak financial results over the last 10–15 years, even without CO_2 tax being levied, and with several years' exemption from electricity charges.

An important precondition in the model's calculations is that Norway achieves an exogenously stipulated export surplus, the size of which can be a critical parameter. The greater the income from the petroleum sector, the less the 'need' will be for other export income. This also means, however, that if the income from petroleum exports fails – for example, because oil prices fall as a result of an international agreement on limitation of CO_2 emissions – then we will meet a greater need for traditional export sectors if this sort of exogenously stipulated surplus is to be achieved (see Mathiesen, 1991).

Alternative regulations

In calculations where the processing industry and international shipping are exempted in the model, these industries will be, naturally enough (in the same way as in scenario 4 in the MSG simulation) only insignificantly affected. The cost to society will be 1.7 per cent, compared with 0.3 per cent (in alternative (b)). The reason is, quite clearly, that there are fewer industries to bear the costs of the emission reduction and that these must reduce emissions even more. The emission-intensive domestic transport sectors and their customers suffer; their level of activity is reduced because of the cost, leading to increase in the price of their services. Road traffic and consumption of petrol and heating fuels is therefore reduced more than if all sectors pay the same tax. If the economy is not optimally adapted with respect to taxation and environmental regulation then, according to economic theory, it will not be necessary for everyone to pay the same tax. Calculations tend to indicate that sectors which face demand with low price elasticity – for example, cement – should pay higher tax. From the government's point of view, not least to be able to withstand lobbyists more easily, an across-the-board tax would be preferable to a sector-specific set of charges.

SFT has drawn up a catalogue of measures for reduction of greenhouse gases. The measures are described in the form of investments in technology and changes in operating costs and emissions for individual sectors. Inclusion of such measures in the model reduces 'demand' for emission rights, reducing the tax and benefiting all sectors. Several of the measures are so comprehensive that the changes in taxation are considerable. The advantage of a general equilibrium model is that it can provide a fairly precise answer as to whether a measure is economically profitable or not.

The greenhouse effect is caused by several of the greenhouse gases, CO$_2$ being the major contributor. An effective regulation would cover all significant emissions. Håkonsen (1993) carried out calculations where the Norwegian emission of greenhouse gases, translated into CO$_2$ equivalents, is reduced. These analyses show that there is no great difference for Norway if CO$_2$ alone is regulated, or if all greenhouse gases are regulated. It is, however, worth noting that while CO$_2$ emissions are common to all sectors and from several different sources, and for this reason are especially suitable for fiscal regulation, the two CF gases are created only in the manufacture of primary aluminium and 40 per cent of the N$_x$O emissions originate from production of nitric acid. In this way, administrative measures, as considered by the Norwegian Pollution Control Authority, would probably be equally as efficient as a tax for these emissions.

Our analyses have shown that emissions of the various gases are significantly complementary. The reduction in CO$_2$ can today take place only through reduced combustion, and the emissions of CO, SO$_2$, NO$_x$ and NMVOC will also be reduced by a regulation of the CO$_2$ emissions (see Rønning, 1994).

19.3   Analysis of adjustments

We have summed up the results per sector from several analyses. Their views of the consequences of reduction on the levels of activity and employment in the sectors on Norwegian CO_2 emissions are in relative agreement. A few sectors are significantly reduced, the others see a small reduction or even a small increase. A characteristic feature of several of those sectors which are seriously weakened – for example, production of ferro-alloys and primary aluminium – is that many companies are located in industrial regions with few alternative employment opportunities. The problems which would have to result from the reduction in activity are not addressed well in the above analyses or in other equivalent models. In this section, we shall modify the description of the job market, its process of change and wage formation and finally use these modifications in concrete analyses of the change over a 10-year period. For a more detailed description of the theory behind these analyses, see Mathiesen (1995).

There seem to be good reasons to presume that in those sections of the employment market with which we are concerned there is unwillingness for employees to move to other areas, lack of willingness or ability among them for retraining, and employees are not very attractive to other employers in other industries. In this way, some of the labour force made idle in one sector will be lost when the employment structure changes. These losses mean lower production opportunities for the economy over a number of years. We must, therefore, assume that some of the labour force does not move, that they remain in the area and that local unemployment figures will rise.

We would now like to model both wage inflexibility and inertia against change. Wage inflexibility will generate unemployment, while change inertia will lead to loss of jobs as a result of a pattern of employment which differs from the original situation – that is, that before the introduction of the CO_2 tax. We shall not attempt to obscure the fact that the speed of such change, and the establishment of new jobs, is a hotly debated issue for economists. It is also a fact that neither economists nor others have good enough models, or sufficient exact empirical insight, to be able to state anything with great certainty. The economic models which are used here are probably more poorly suited to react to the establishment of new industry than they are to react to change in (and especially reduction of) already existing industry.

If we are going to come to grips with the problems involved in a 20 per cent reduction in CO$_2$ emissions compared to the reference path, we must segment the labour market in the model. It is not clear which criteria we should use to segment the market; possibilities include the type of labour (employees' skills), type of industry (here called a sector) and companies' location. We have chosen to group in five categories, according to how strongly the industry is affected. The distribution is based on a review of previous analyses, which also operate with an undivided labour market and one wage.

We follow up the analyses in this chapter, and model endogenous availability of labour and a situation where the revenue from the CO$_2$ tax is used to reduce employers' contributions. In other words, we are examining a situation where the goal is a 20 per cent reduction in total Norwegian CO$_2$ emissions compared to the reference path in 2000, and we use the following alternative preconditions in the labour market on the following terms:

1. Ample vs. limited opportunity for structural adjustment.
2. No wage restriction vs. a 'tight' wage restriction for the two most exposed groups. By this we mean that wages in these two sectors must not fall below 95 per cent and 98 per cent of the average wage, respectively. We allow the greatest fall in the sectors (refineries and production of ferro-alloys) where the impact on the level of activity is greatest. We do not fix a lower limit on the other sectors.

With two results from each relation we get four scenarios. We designate these scenarios with the letters AA–DD (to prevent misunderstanding) – that is, AA represents major, although imperfect structural adaptation opportunities and no wage rigidity, while CC represents little opportunity for adjustment with simultaneous wage rigidity. As we have already said, the latter combination will provoke overt unemployment.

Figure 19.5 shows the following computed indicators for the four corresponding equivalents:

Alternative AA: major adaptation and flexible wages
Alternative BB: little adaptation and flexible wages
Alternative CC: little adaptation and restrictive wage policy
Alternative DD: major adaptation and restrictive wage policy.

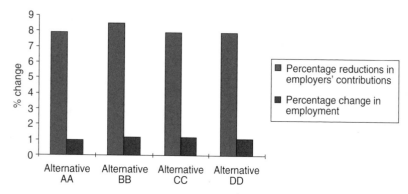

Figure 19.5   Macro-indicators in four scenarios for the employment market

'Reductions in employers' contributions' is defined as the reduction in the number of percentage points in employers' contributions as a result of the CO_2 revenues, 'change in employment' is defined as the percentage change in total employment in the alternative.

We note that there are only minor variations in the macro-indicators of the four scenarios. This is because, as we have previously remarked, the total significance on the economy of the sectors is limited. BB, which represents little adaptation and flexible wages, does differ somewhat from the other three. BB ties the labour in the 'problem sectors' and allows significant wage reductions (see Figure 19.5). The model forecasts a certain substitution in the direction of more investment in labour, and therefore slightly higher production (than in AA, CC and DD) in these sectors. For this reason, the 'pressure' increases on the CO_2 goal, and the tax rises slightly, which increases the revenues gathered and allows a greater reduction in employers' contributions. We can see, however, that these effects have a negligible impact on total employment.

Alternative AA is the one which most closely approaches the previous model analyses, showing perfect adaptation in an unsegmented labour market and one wage for the homogeneous workforce. Alternative CC, which has especially slow adaptation and wage restrictions, is perhaps the most realistic alternative within our 8–10-year analysis horizon. In the even longer term, as in our MSG scenarios in Chapter 18, workers who have been laid off in this scenario reach retirement age and therefore are omitted from our analysis; they are replaced in the workforce by younger workers. Given this interpretation, we can estimate the underestimation of earlier welfare forecasts for this sort of structural adaptation to be in the order of 0.1 per cent, which is also the difference in the welfare index for AA and CC.

Let us now study the fluctuations in the employment market in greater detail. Category 1 (refineries and production of ferro-alloys) will be the hardest hit, suffering job losses of almost 50 per cent in three of the scenarios. This is as in the previous analyses. Only in alternative BB, where there is little change to other types of employment, and where all the 'locked-in' workers share available work in these two fields, is there less loss of jobs.

We also can see, however, that the work-sharing leads to a wage reduction of, in all, 65 per cent. It might prove to be that this solution would not be put into place in the market on a permanent basis, as is inherent in equilibrium. In addition, even if the concept of public subsidy could be entertained, in order for the employees' total income to be more or less maintained, the description of the production is probably misleading. As we have pointed out above, as a result of heavily decreased wages, labour is replaced by other factor investment. Not much, certainly, although probably more than the realities would indicate in the two relevant fields. The processing industry is capital-intensive, and in all of the scenarios we study, earnings do not even approach ordinary depreciation on capital in category 1. Even though more labour-intensive technologies could be found, these industries would not be able to undertake the necessary investment, nor would they probably be profitable.

We have indicated that alternative CC is perhaps the most realistic. The wage restrictions are binding in this alternative, and open unemployment occurs, in contrast to alternative BB where there is hidden unemployment because of work-sharing. Employment in categories 1 and 2 in CC are, to all practical purposes, the same as in AA thus also the same as we have found in earlier analyses. The difference between CC and DD is the number of workers who will move and thereby find other employment. We note that total employment in Figure 19.5 is therefore somewhat higher in DD than in CC.

If we examine the individual sectors in more detail, the model reveals further details. Changes in the level of activity for the sectors is, however, with few and minor exceptions, surprisingly similar in all four scenarios. Once again, alternative BB distinguishes itself. Here we saw that the reduction in employers' contributions was greatest. This combination hits cement production, while the other sectors emerge somewhat better. Totally speaking, all industries emerge poorly from this sort of fiscal policy change. It is interesting that the changes in the levels of activity in the sectors in scenarios AA and CC are as good as equal. This means that the errors made on earlier analyses with regard to the fluctuation for individual sectors are also small.

The analysis of the change has provided us with little new material, in the sense of new results. On the contrary, our concept of the consequences, reached on the basis of earlier analyses, has been confirmed. A regulation of Norwegian CO_2 emissions through equal taxation over all sectors would mean little for the Norwegian economy, but it hits some sectors very hard – that is, the emission-intensive and probably the export-oriented sectors and their stakeholders. This confirmation is, however, also a useful result. It enables us to be more certain about the conclusion that it is robust with regard to the parameter change which has been analysed.

Is there, however, anything in our analysis and model which indicates that the results, and the knowledge which this report intends to provide, is wrong? In the first place, let us assume that the premises about the Norwegian CO_2 goal, and its implementation using an equal CO_2 tax, are correct. Could the description of the sectors be wrong, so that we do not include all sides of their activity and make errors concerning the consequences of the CO_2 tax? We have indicated two conditions: export demand and capital return. We have pointed out that the opportunity for costs transfer in the global market may well be less than we believe – that is, that demand may be more price-elastic. The consequences of more elastic demand are that those industries to which this applies would experience more rapid recession. This recession would, however, relieve the CO_2 problem, and ease the situation for everyone.

The low profitability for the processing industry and the other emission-intensive activities give, according to the model, little or no return on capital. We have implicitly assumed that man-made capital has zero opportunity cost, and furthermore, that it does not require maintenance. The man-made capital of the processing industry has hardly any alternative use, so that the opportunity value is actually zero. On the other hand, if equipment is to be kept operative, routine maintenance and modernisation do cost real money. To the extent that there is not sufficient profitability, and the creditors are not willing to renounce their interest demands, it may not be possible to maintain operation. In that case, the recession would be more rapid than indicated in our analyses.

It would seem today that the most significant weakness in the analysis can be found in the premises which surround a unilateral

Norwegian reduction! The government has signalled that it will not be able to fulfil its former statements about a stabilisation of Norwegian emissions in 2000 at 1989 levels. This is probably because of concern for the major consequences for what is, according to our analyses, a very small minority (of workers, managers and capital holders), but who are very well organised and very influential. Mathiesen (1991) has analysed the consequences of both unilateral measures and participation in an international agreement and concluded:

> also in the agreement scenario, where Norway follows up an international agreement about the stabilisation of emissions, the activity in parts of [energy- and emission-intensive] ... industry becomes reduced. The reasons are several and interwoven. The immediate reason is that it is mandatory for all production to take into account that disposal of CO_2 into the atmosphere is no longer free of charge. This leads to a cost which industries in this sector are financially unable to meet. Most industries export to the global market in exchange for currency. In the agreement scenario, where revenues from oil are assumed to fall, there will be increased need for foreign currency. This needs both increased export and reduced imports. The reason that the export industries are still limited in their extent, is partly because their customers are unwilling to pay the increased costs in full, and partly because not all potential export is profitable for Norway. We can, in fact, use our resources with greater return in other activities, for example, in import reduction. The model shows that production of ferro-alloys, fertilisers and petrochemical products suffer most. These industries are fairly unprofitable even today, when no CO_2 tax is levied. Profitability for society will hardly be greater under a future regime of international CO_2 agreements.

The computations are not repeated with the extended model which has been used here. Our computations with various price elasticities in export demand, and therefore with various degrees of cost transfer in the global market, show that the 'picture' we have drawn is mainly unaffected.

20
Agriculture, the Environment and Employment[*]

20.1 Alternative development courses for Norwegian agriculture

Two future scenarios based on the JORDMOD model are compared with a baseline solution. The two new equilibria are assumed to exist 25 years in the future, and show two possible interpretations of the signals given in the White Paper, 'Proposal to the Norwegian Parliament, 8 (1992–3)'. They are labelled 'the GATT alternative' and 'further liberalisation'. 'Northern' agriculture includes the regions which were inside the area in the agreement which resulted from the membership negotiations with the EU. The 'northern' agriculture region includes more than half of the agricultural districts in Norway. The rest of the agricultural area is defined as 'southern' agriculture. A map showing the division into northern and southern agriculture can be found in various works, including Børve *et al.* (1994, p. 118).

The 'GATT alternative'

The GATT alternative is based on Norway's obligations under the General Agreement on Tariffs and Trade, and on an interpretation of the signals given in White Paper, 8 (1992–3). Import prohibitions are lifted, and a system of fixed (customs) duties and import quotas is imposed. There is also a movement from price subsidies to land subsidy. Total agricultural subsidies will be reduced, although continued high goals are assumed with regard to employment and land use. The ambition is for 55 000 man-years in the agricultural sector, 40 000 of which are to be in the northern region. In addition, the land

---

[*] This chapter is mainly built on Rickertsen, Holand and Rystad (1995) and Lothe (1995).

Table 20.1 Maximum permissible farm sizes in the scenarios

	Baseline solution Country	GATT alternative		Further liberalisation	
		Northern	Southern	Northern	Southern
Dairy cattle	15	20	30	30	100
Suckling cattle	–	56	56	–	56
Goats	70	100	100	70	–
Bulls	40	40	100	100	100
Sheep	50	100	200	100	500
Sows	21	35	100	70	100
Poultry	2 000	6 000	10 000	10 000	10 000
Grain (hectares)	300	300	1 000	600	1 500
Potatoes (hectares)	100	100	150	100	150

Source: Mittenzwei, Huus and Prestegard (1994, p. 28) and Børve *et al.* (1994, p. 61).

requirement is for 6.2 million hectares and the requirement is for grain production at a level of at least two-thirds of present production in the southern region. The political reality behind such demands can, of course, be debated, although it is possible to find measures which will allow us to achieve these milestones. A further demand is for farm sizes not to exceed those shown in Table 20.1. Milk quotas are reduced from 1800 million kilos in the baseline solution to 1650 million. Costs in the agro-processing industry are assumed to have been reduced by 15 per cent.

'Further liberalisation'

This alternative contains elements of adaptation to EU norms. The main deviations are connected with heavy subsidies for southern agriculture, although the maximum subsidy rates to northern agriculture are not fully exploited. This enables a two-way split in Norwegian agriculture to be avoided. The alternative stays within the framework of Norway's obligations under the GATT agreement. The scenario is based on an interpretation of White Paper, 8 (1992–3) which de-emphasises employment in Northern agriculture compared with the GATT alternative above. In this alternative, parity with Danish price levels is assumed. Import of agricultural produce takes place at Danish prices *plus* transport costs to Oslo. Table 20.2 shows that the prices of pork, poultry, feed corn, concentrates and eggs are more than halved.

Table 20.2   Wholesale prices in Norway and Denmark, per kilo or per litre, 1992 (Danish prices include transport costs to Oslo)

	Norwegian (NOK)	Danish (NOK)
Cheese	48.65	30.14
Butter	23.37	27.29
Drinking milk	7.89	4.88
Beef	34.77	20.71
Pork	32.85	11.92
Poultry	22.84	5.43
Feed grain (barley)	2.28	1.16
Grain (wheat)	2.67	1.12
Concentrates (cattle feed A)	3.49	1.58
Potatoes	2.38	1.20
Eggs	18.39	8.32

Source: Mittenzwei, Huus and Prestegard (1994, p. 23).

This low-price scenario reduces any problems with overproduction, reduces any trade disagreements with other countries, increases efficiency in agriculture and gives consumers cheaper food, possibly even reducing certain local environmental problems. The low-price scenario does, however, also create problems. Producers' incomes are very low, production may be displaced from one region to another and cause problems, production of certain public goods may fall too low and other environmental problems can increase.

Problems which will arise in the liberalisation alternative are most easily reduced by subsidising production of public goods, supporting environmentally sound production methods and possibly through direct income support. One approach to these sorts of subsidies is granting subsidies per head of stock or by unit of land. Subsidies can be differentiated by region, production volume and production methods.

In this alternative, producers receive compensation from regionally differentiated subsidies. The subsidies are, however, not differentiated in terms of production volume or production methods. They are designed to produce a result with roughly the same production figures and distribution as in the GATT alternative. This means that the scenario aims at maintaining production in all parts of the country, although the method is cheaper than in the GATT alternative. The scenario allows for larger farm units in southern agriculture than we saw in the GATT alternative (see Table 20.1).

20.2 Equilibrium solutions in the long term with JORDMOD

Table 20.3 shows production, employment, land use, capital, welfare surpluses, agricultural subsidies and average farm unit sizes in the baseline solution (Børve *et al.* 1994) and the two scenarios. Four main tendencies emerge: total agricultural production shows no dramatic change, production factor use (especially labour) falls, protective subsidies decrease and the average farm size increases, often considerably.

The reason why the production effects in the alternative for further liberalisation are as moderate as they are is that this scenario means low consumer prices which, in turn, lead to increased demand for Norwegian agricultural produce.

The results represent in many ways the 'worst-case scenario'. This is because the costs of change are overestimated in the model because of rigidity on the production side. Production should, therefore, be maintained with a lower budget subsidy than that shown in Table 20.3, without reducing employment further. We should, however, remember that many of the results are driven by direct barriers and the minimum requirements of the scenarios.

The model indicates a relatively heavy reduction of grain production in the GATT alternative. Production is still, however, maintained at two-thirds of present levels, as this is a specified requirement of the scenario. Grain production fares better in the second scenario because of the high land subsidy. This indicates that it is possible to achieve a fairly high grain production through a combination of relatively low prices and high land subsidies. Milk production approaches the market balance in both scenarios; this is a result of the barriers on milk production stipulated in the design of the scenarios. Production of beef, mutton, feed grain and potatoes remains fairly constant.

Pork and egg production increases in the scenario for further liberalisation. The increase is a result of the extremely high subsidies per pig and chicken. These rates could seem to be unrealistically high, although they could be reduced heavily if a price increase of a couple of kroner per kilo on pork and eggs is permitted. This will reduce the budgetary subsidy, but increase the protective subsidy to a certain extent. The subsidy per pig and chicken can also be reduced because production is primarily high in both scenarios, at the same time as the model underestimates the opportunities for change.

The budgetary subsidy is relatively constant in the two scenarios and the baseline solution. The protective subsidy falls heavily, however,

Table 20.3 Production, factor consumption, subsidy and average farm unit size

Production (million kilos/litres)	Baseline solution		GATT alternative		Further liberalisation	
	Southern	Northern	Southern	Northern	Southern	Northern
Milk	659	1 141	472	1 178	631	1 035
Beef	21	60	31	50	51	32
Pork	67	17	82	8	113	0
Mutton	5	20	0	22	0	25
Egg	52	0	43	10	51	9
Potatoes	354	4	250	148	290	68
Food grain	149	0	165	0	136	0
Feed corn	922	101	575	60	775	101
Employment (1 000 jobs)	28	42	15	40	14	23
Land use (1 000 hectares)	3 700	3 600	2 600	3 600	4 500	2 300
	Total		**Total**		**Total**	
Capital (1 000 million)	60		49		41	
Welfare surplus (1 000 million)	14		18		22	
Budgetary subsidy (1 000 million)	10		9		10	
Protective subsidy (1 000 million)	9		6		3	
Farm size						
Milk (year cattle)	10		17		33	
Pigs (sows)	21		39		100	
Sheep (winter fed)	50		50		91	
Eggs (chickens)	2 000		2 000		9 430	
Grain (hectares)	300		600		676	

because product prices fall. The total subsidy is then reduced by NOK 3 and 6 billion, respectively in the two scenarios.

Several regional distortions have taken place, but it must be remembered that these are uncertain. Production of milk, food grain and potatoes in the GATT alternative is heavily reduced in southern agriculture. This is because this alternative stipulates an explicit barrier to maintain high employment in northern agriculture. The reduction in production also leads to heavy reduction in land use in southern agriculture. In the same way, potato-growing increases appreciably in northern agriculture. In the alternative for further liberalisation, production of pork and beef increases in southern agriculture, while feed corn production falls. In northern agriculture, potato production, in particular, increases, while beef and pork production decrease.

The model fixes detailed requirements for return on capital and labour per type of farm unit. If prices and subsidy rates are fixed so high that the return demands are met, the model 'forces' farms to produce at about their production capacity. If the subsidy rates are set so low that the demand on profitability is not met, then the farms no longer produce.

Changes in production lead to changed factor consumption. Total land in use is reduced by between 10 and 15 per cent. The reduction is greatest in southern agriculture in the GATT alternative. Land use in southern agriculture seems exaggeratedly high in the further liberalisation alternative. Land in use increases by 20 per cent without major changes in total production. This is because the very high land requirements make suckling cow production profitable (given the subsidy rates built into the scenario). The increase in land use seems to be unrealistically high, although the result indicates that relatively large areas can still be devoted to agricultural land.

Employment suffers apparently heavy reductions in both alternatives. If we assume that establishment of a new long-term equilibrium will take 25 years (after agriculture has lost the baseline equilibrium as a result of policy change), then an annual reduction of 2.5 per cent (the trend in the post-war period) in 25 years will lead to a decline from 70 000 jobs (baseline solution) to 37 000 jobs (further liberalisation). This scenario can also be interpreted so that it means a reduction in labour consumption which is equivalent to the historical trend in the post-war period. The model also takes partly into account the process of technological change when production, as a result of deregulation, moves from relatively inefficient use to more efficient use. New technology, however, develops over time, as does more efficient machinery, better grain and grass types, and so on. Over time, the access to new technology will lead to labour reductions in agriculture, whether deregulation takes place or not.

These sorts of technologically induced job losses will take place in addition to the policy induced job losses. There is, therefore, reason to believe that the actual job losses in the 20–30 years following deregulation will be greater than the effects which the model computes. Fewer jobs are lost in the 'GATT alternative'. The relatively high job figures in this scenario are a result of the minimum barrier of 40 000 jobs in northern agriculture.

Change from the baseline solution to further liberalisation means a reduction of somewhat more than 30 000 jobs and a subsidy reduction of around NOK 6000 million. According to the model computations, it will cost around NOK 200 000 annually to maintain each of these jobs. The capital base of agriculture is reduced in both scenarios.

Average farm size increases as production is relatively constant, while jobs are lost. This applies especially in the production of milk, sows, eggs and grain in the scenario for further liberalisation. Production will take place in larger and more efficient farm units than we see today. Inherent in the alternative is a movement towards Danish agricultural conditions, where the average dairy herd was 36 head in 1991. The barrier to farm size prevents the model from predicting even larger farm units. In other words, there is reason to believe that the structural change and job losses in an equivalent restructuring of the framework conditions for agriculture (in time) would be even greater than the model results indicate.

The degree of specialisation in the model is high. The only mixed production which is built into the model is pig-farming and grain, egg production *plus* grain, and potatoes *plus* grain. The model cannot, therefore, be applied to consider whether deregulation would lead to increased or decreased agricultural specialisation.

20.3 Simulation of a gradual change to lower producer prices with NAP

NAP (Norwegian Agricultural Policy), which is a simulation model for Norwegian agriculture, has been applied to simulate the effect of a gradual reduction of prices to Danish producer levels. This is a reduction which is expected to take place between 1991 and 2001. For a variety of reasons, the results are uncertain. One main source of uncertainty is that the model is used to study the effects of major policy change. The data in the model is based on historical data from the period between 1959 and 1986. In this period, prices were never as low as today's Danish prices. This source of uncertainty is something that this model shares with MODAG/MSG as they are applied in the Project for a Sustainable Economy.

The simulation finds that total grain and potato land increases by 2 per cent. The area under wheat in 2001 is as large as it was in 1987. This indicates that grain production can be maintained, in the short term, at lower prices than today. Production of mutton and eggs increases despite the heavy price reductions. This increase, which is partly a result of trends, seems exaggerated. Production of beef, bacon, carrots and tomatoes is reduced as a result of the lower prices. Dairy production is regulated by quotas, and no range has been estimated for this. Meat from cows and veal are treated as dairy by-products in the model. Production will therefore fall as a result of the fall in milk production.

20.4   Resource use following deregulation

Our attempt at a systematic view of changes in resource use in a deregulated agriculture uses as its basic data the specialist milk and grain farms from the annual 'farm surveys'. As figures vary widely from year to year we have chosen to work from mean figures for 1991, 1992 and 1993.

Specialised dairy production

The most obvious characteristic of this resource-use data is the great variation from farm to farm. Consumption of chemical fertiliser varies from NOK 0.10 to 0.50 per litre at production levels of around 50 000 litres of milk in northern agriculture. The variation in the use of pesticides is also great. In southern agriculture costs vary from NOK 0 to 0.08 per litre of milk. Variations in machinery costs, medication and fuel show comparable size. This indicates that there is considerable variation in production methods from farm to farm, although the amounts produced are roughly the same. The range in variation makes it difficult to find statistically significant differences in resource use between the various groups.

Considerable differences exist. Can these differences be explained by generally poorer production conditions for small farms, or by small farms buying relatively less fodder and/or using relatively more of their production factors to produce such other agricultural produce as meat and potatoes? Production of beef per litre of milk is actually least for the largest farms, but this is partially compensated for by the fact that other production income per litre of milk is greatest in this category. The amount of commercial feed decreases slightly, compared to the size of production. Nor can differences be explained by variations in natural production conditions, measured in the average milk zone.

Costs fall if we go from the farm size group 11–15 head of cattle to the group 16–20 head, even if the latter group is over represented in the least favourable production areas.

In other words, it would seem that there are certain significant cost differences between groups of farms of different sizes and between farms in different regions. The economic analysis yields several results which are statistically interesting:

1. More and more pesticides per unit of production in southern than in northern agriculture. This, as we have already noted, is because the average southern farm produces more arable produce than the northern farm.
2. More pesticide is used within the large farm group in southern agriculture. The high consumption is within the 16–20 head group, which is somewhat difficult to explain.
3. Less machinery (in total and in northern agriculture), fertiliser (in southern agriculture) and fuel (in total and in northern agriculture) is used on large farms than small farms. This indicates more effective use of these resources on larger rather than on smaller farms.

Reduced prices lead to reduced milk production. In the two JORDMOD scenarios we have analysed production falls to a domestic market balance. In this way, the loss-bringing export of dairy products is heavily reduced. In an isolated sense, this leads to reduced factor consumption, including polluting and non-renewable resources in the Norwegian agricultural sector.

The most important changes in resource use are linked to changes in the average size of farms. In the two deregulation scenarios the average size of the dairy herd increases heavily compared with the present day. This leads to reduced labour use and reduced costs in dairy production. Use of machinery, fuels and chemical fertilisers per litre of milk produced will also be reduced. These differences between large and small farms are statistically significant for machinery and fuel. As these differences cannot be explained by variations in the control variables, we can conclude that a change to larger production units in dairy production will lead to a reduction in use of these production factors per unit produced. Use of pesticide and medication per litre of milk is, however, only marginally affected by the extent of production.

The regional changes in dairy production are not unambiguous in the two scenarios. It is, therefore, not possible to say anything certain about the consequences any regional displacement production will have on resource use per unit produced in Norwegian dairy production. Based on the analysis of the cost structure in Norwegian dairy production, we can conclude that there does not seem to be any conflict between economically more efficient dairy production and use of non-renewable and heavily polluting resources. On the contrary, a deregulation of dairy production will probably lead to reduced use of such production factors per unit produced.

Specialised grain production in the lowland districts of eastern Norway

Farming surveys show, on specialised grain farms, that consumption of artificial fertilisers, pesticides and machine costs vary heavily with the same production volume. This tends to indicate great variation in production methods for different farms, even when the volume produced is around the same. The variation in use of pesticides and fuel between the various groups is, however, small. Machine and fertiliser costs are fairly heavy, while pesticide and fuel costs are far less. There are fairly small variations between the different size groups.

The most conspicuous resource group effect of deregulation is reduced labour use resulting from larger farm units. According to the two scenarios carried out using the JORDMOD model, the average area of grain per farm unit will double compared to today's levels. This will probably have small consequences for use of machinery, chemical fertiliser, pesticides and fuel per grain unit produced. To the extent that there is any effect, it is not negative. Average costs per unit produced are lower on the large farms than on the small for all cost categories. None of the differences in our analysis of the farm survey data are, however, statistically significant, so the conclusions must be regarded with significant reservations.

20.5   Norwegian agriculture under co-ordinated European handling of environmental problems in dairy and animal production

Nitrate pollution from agriculture in Norway and the EU

In 1987 more than twice as much nitrogen per hectare was released in the Benelux countries compared with Norway. When the density of farm animals in Belgium and the Netherlands is taken into account, this leads to a far higher pollution level per hectare agricultural land than in Norway (Simonsen, Rystad and Christoffersen, 1992). The conclusion of the report from the third North Sea Conference in 1991 about managing the nitrate emission is that discharges of nitrate salts into the North Sea from Norwegian agriculture are very small when compared with equivalent discharges from the Netherlands, the UK and Germany. This is a result of the low level of agricultural production in Norway, as well as an intensity so low that only a small proportion of the discharges reaches the sea.

Both the Netherlands, northern Italy, Belgium, France, parts of Germany and parts of southern England struggle with the problem of high levels of nitrogen in drinking water. Denmark has reduced its discharges considerably through its domestic measures. Most nitrous discharges in these problem areas come from animal farming, a result of the high density of animals in these districts. In 1991, a new EU Directive was adopted, which fixes actual limits for the nitrate levels in drinking water. The Directive requires that each country must identify problem areas within two years. Countries are then allowed a further two years to prepare suitable measures to reduce the nitrate levels. These programmes will be implemented over a four-year period, meaning that it will take eight years from start until the goals of the directive are reached.

Agricultural policy and pollution

The EU common agricultural policy (CAP) has to a great extent contributed to high pollution levels in certain EU regions. A high degree of self-sufficiency, and at the same time high farm incomes, have been given priority at the cost of public goods in the community, and especially at the expense of the environment. Local competition and efficiency through protective measures have been more important than sustainable agricultural methods. Attempts have been made to remedy this using comprehensive changes to agricultural policy. In addition to the nitrate and drinking water Directives, the MacSharry Plan (Box 20.1) will lead to major change in agricultural adaptation.

Box 20.1   The MacSharry Plan (1991)

The MacSharry Plan (Commission of the European Communities, 1991) aims at separating subsidies to producers from product prices, and when combined with allowing land to lie fallow, will also contribute to a reduction in nitrogen emissions. MacSharry was the Irish representative in Delors' second EU Commission.

Production quotas for milk have, to a certain extent, already reduced the number of animals. The number of milk cows has been reduced by between 12.5 per cent and 18 per cent in various EU countries from 1986 to 1991 (Leuck, 1994). The effect that this has had on pollution levels has varied, as milk quotas in individual EU countries are marketable, so that certain regions in the individual country can have very high density of animal population even when the total number of milking cows is reduced. Technological change in many EU countries has been rapid, despite milk quotas. In Denmark, the economies of scale in dairy farming have been exploited so that there are now fewer dairy farmers and higher production levels per farm than before (Jervell, 1994).

Farmers in Norway have also benefited from price subsidies and other production-promoting subsidies, although we have not suffered the same pollution problems as certain EU regions. Quota regulation, limiting numbers of cows, as well as the traditional family farm structure have contributed to milk being produced on small farming units, and the producers have had little or no incentive to exploit economies of scale in dairy farming. The proposal from the Agriculture Minister for the introduction of marketable quotas within the regions (*Aftenposten*, 20 December 1994) would provide the most efficient producers with incentives to increase production, while marginal producers would close down or reduce the extent of their production. This does not necessarily have to lead to pollution, as Norwegian animal farmers are obliged by law to have certain minimum areas of rough fodder land per animal. As we have seen earlier, there is no reason to believe that heavily stocked farms would use more fertiliser than smaller farms.

Reduced nitrate emissions and impacts on trade in the EU

Norwegian dairy production makes up around 2 per cent of dairy production in the EU. The Norwegian dairy sector is, therefore, facing given prices in the global market, while major changes in production within the EU and export/import can affect global price levels. For this reason, it is natural to examine the EU's proposed environmental measures and these measures' impact on global markets first and then analyse the competition situation for Norway, given these changes.

Belgium, the Netherlands and Denmark must reduce residual nitrogen levels if they are to satisfy EU's Nitrate Directive. Analysis shows that this will lead to heavily reduced export for several products, and in certain situations a change from net exporter to net importer. Both

Belgium and the Netherlands changed from being net exporters to being net importers of butter and the Netherlands cut cheese exports down almost to a self-sufficiency level. Belgium experienced major cuts in exports of dried milk. In time, the Netherlands will become a net importer of consumer milk. Denmark, on the other hand, will become a net exporter of dairy products, even after meeting the goals of the nitrate directive.

Haley (1994) analyses three different policy scenarios for reduction of agricultural pollution:

1. the MacSharry Plan
2. the Nitrate Directive
3. taxing commercial fertiliser.

The impact on trade is analysed using the SWOPSIM model (Static World Policy Simulation, see Roningen, Sullivan and Dixit, 1991). The model has, however, a high aggregation level, which makes it fairly unsuitable for detailed policy analysis.

The MacSharry Plan reduces export of butter, dried milk, and beef, while cheese exports increase. Prices on the world market of butter and beef increase substantially. Lower grain prices in the EU lead to a heavy upswing in pork production, compensating for the downturn in beef production. Increased demand in the EU for grain means that grain export falls heavily, while maize corn import increases by more than 300 per cent. This leads to strong increases in grain and maize prices on the world market.

If the MacSharry Plan is combined with the Nitrate Directive the model result changes. The Nitrate Directive leads to reductions in animal density, having special consequences for butter, cheese and meat production. Internal grain demand is reduced and the net export of wheat, maize and other grain types is reduced less than in the case of introduction of the MacSharry Plan alone. The reduction in exports of butter, cheese, dried milk and beef leads to higher prices on the world market for these products. A 75 per cent tax on commercial fertilisers would have a relatively small impact in trade, compared with the other scenarios.

If the efficiency in reduction of residual nitrates is viewed uncritically, a 75 per cent tax on commercial fertilisers would be more efficient than the MacSharry Plan or the MacSharry Plan combined with the Nitrate Directive. The weakness of this tax is that it is uniform, and that the nitrate concentration will, therefore, not be reduced sufficiently in those areas defined as threatened. Reduction in use of commercial fertilisers will probably take place in areas of intensive cultivation and, unless these areas are near districts of intensive animal farming, the impact on the environment will be negligible (Haley, 1994).

The consumer milk market has not been included in Leuck's and Haley's scenarios. Areas which are defined as most polluting or threatened are, however, also areas with the greatest potential for cheese and butter production, and it would seem probable that these areas will increase demands for import of milk for processing if their own dairy production decreases. In the long term, it should be possible for expansion of dairy production in areas which are not, at present, defined as polluted and these areas will export milk for processing in the more threatened areas. Price increases on the global market could also lead to new regions choosing to invest in dairy production and export of consumer milk or processed products, including butter and cheese.

The consequences for Norwegian dairy production of liberalised markets and environmental regulation

The most competitive regions in the EU show both high exports of food produce and also pollute the most. Norway is not particularly competitive in food markets, nor does it experience major problems with nitrate pollution. To illustrate this problem, we can use a partial analysis (for example, of the cheese market) where Norway would be a net importer if there was a free market.

Assume to begin with that Norway has internalised the environmental costs of production in the supply curve, while the EU supply curve reflects only the costs of production. It may be assumed that the rest of the global market has no problem with polluting production. An internalising of environmental costs in the EU would lead to reduced EU export, and a global price rise (equal to the introduction of the EU's Nitrate Directive). Norway is only a minor player in the global market and for this reason is not able to influence the price of cheese. Norway therefore faces an export supply curve which is perfectly elastic (horizontal).

In the change from self-sufficiency to free trade Norway, coming into a situation where the exporting country does not internalise environmental costs, would reduce cheese production form Q to Q_m. This leads to a heavy reduction in producer surpluses. If, on the other hand, exporting countries correct for the deterioration in the environment caused by cheese production, then Norway would become more

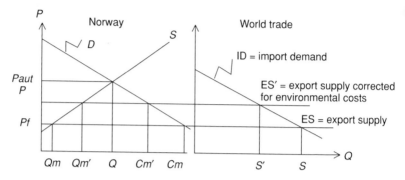

Figure 20.1   The effect of liberalised markets with and without internalised pollution costs in export markets

competitive and only reduce production to $Q_{m'}$. The free trade scenario will thus seem less drastic from the Norwegian producers' point of view. Similar scenarios can be drawn up for production of butter, dried milk and beef, which all increase in price on the world market if the EU introduces more stringent environmental regulation (Leuck, 1994; Haley, 1994).

The scenario has not taken into account how a liberalisation of trade influences the supply curve of Norwegian producers. More factors than just the price of the finished product are significant when the production decision is made, and it is important to know what determines the supply when assessing the whole picture of how the changed measures have affected trade. We will now see what determines the supply of milk, which is the decisive production factor in cheese, butter and dried milk production, and how various liberalisation scenarios impact on dairy farmers' competitive edge.

Annual time-series data from 1959 to 1992 have been used to estimate the aggregate supply of milk in Norway (Budsjettnemnda for jordbruket (Budgetary Committee for Agriculture), 1985, 1993). In the quotaless period between 1959 and 1982, the estimated price elasticity for milk is 0.29 (short-term) and 0.34 (long-term). This means that an increase in milk prices of 1 per cent will lead to an increase in production of 0.29 per cent in the short term (or the opposite in the case of a price reduction). The corresponding price elasticity for concentrates is –0.20 (short-term) and –0.24 (long-term). If we examine the whole period between 1959 and 1992, the price elasticity for milk falls to 0.16 (short-term) and 0.22 (long-term), while the price elasticity for concentrates remains –0.20 (short-term) and –0.24 (long-term). These changes in adaptation seem natural when quota regulation limits the opportunities for the producer to increase production through a rise in the price of milk.

We will make assumptions here which are consistent with the further liberalisation scenario, where import of agricultural produce occurs at Danish price parity *plus* the cost of transport to Oslo. In this scenario, milk prices fall from NOK 7.89 per kilo to NOK 4.88 per kilo. A reduction of this nature would make quota regulation unnecessary, and the price elasticity from the free trade period (1959–82) would therefore be representative for adaptation by the producers to the price changes.

A liberalisation of trade, where only the MacSharry Plan is implemented, can be compared with a liberalisation scenario where the Nitrate Directive is also implemented. The main claim is that Norwegian producers emerge relatively stronger from the competition the more stringent the EU environmental regulation is. The basis of the claim is that more stringent regulation blocks the polluting EU regions from gaining a competitive edge from the difference between the private and the fully internalising socio–economic comparative advantage.

Percentage-wise prices for dried milk, cheese and butter increase more in tune with the stringency of the regulation. Increased prices on products for which milk is a production factor will, in an isolated sense, contribute to higher milk production. It is very difficult to quantify how large the effect of increased global market prices of cheese, dried milk and butter will be on milk prices. A quantification of this increased demand is virtually impossible because both production and export of these manufactured products has been strictly regulated for a very long time. For this reason, there are no estimated supply functions for these products, or estimates for the industry's demand for milk. It is, however, possible to find an approach to changes in export prices of milk by looking at how much milk is required to produce a given quantity of cheese (butter is regarded as a by-product of cheese manufacture) and use this to estimate changes in milk prices, given the percentage changes in cheese and butter prices (Rickertsen, 1989). If this link is used to estimate changes in export prices for milk, given the various scenarios for regulation of nitrate pollution in the EU, the export prices for milk will increase by 6.2 per cent if the Nitrate Directive is implemented in addition to the MacSharry Plan. If it is assumed that producer prices increase by the same percentage (which they will in a free trade situation where Norway cannot influence milk prices), then milk prices will increase by 1.8 per cent (short-term) and 2.1 per cent (long-term).

These results indicate that Norwegian dairy producers would be, relatively speaking, more competitive if the EU introduces more stringent environmental regulation.

20.6   The consequences of deregulation for agricultural pollution

A decrease in fertiliser levels would, taken in isolation, lead to reduced pollution levels from agriculture. There does seem to be an empirical foundation for the view that N runoff, and to a certain extent P runoff, increases with increased levels of N and P per acre. There is, however, considerable uncertainty connected to the runoff functions. The N runoff varies highly from year to year, all depending on weather and precipitation conditions in the growth season. As the empirical material is relatively sparse, especially for high fertiliser levels, a considerable extent of 'qualified guesswork' has been necessary, and Holm (1989, p. 34) sums up our present knowledge of the link between fertilisation and runoff.

The effects of reduced agricultural prices on N runoff which are shown in Table 20.4 were first presented in NOU 1991: 2B. No new material has become available which seems more reliable.

In the so-called 'GATT alternative', we have introduced actual rates for land subsidy. The model has then calculated how high the minimum prices must be if they are to satisfy the agricultural policy goals represented by the specified demands to employment, total acreage and grain acreage. We found that the average price reduction on grain in the GATT alternative can be fixed at 15–16 per cent. The equilibrium prices in the JORDMOD model calculated in the 'GATT alternative' would, however, increase if we had used a lower acreage subsidy as our basis,

Table 20.4   Reduction in N runoff, per hectare, as a result of reduced product prices

Reduced product prices (%)	Reduction in N runoff per hectare in grain production (%)	Reduction in N runoff per hectare in grass production (%)
15	3	11
30	7	24
45	13	40

Source: NOU 1991, 21B.

and decrease if we had used higher subsidy. The agricultural policy goals which were used as a basis for calculation for the 'GATT alternative' can, in other words, be attained at both higher and lower grain prices than those referred to by adjusting acreage subsidies.

Using the calculations of the Alstadheim Committee (1990) appointed by the Parliament as a basis, the 'GATT alternative', which postulates a reduction in the price of grain of 15 per cent, as well as an equivalent reduction in the unit value of rough feed (which within certain constraints is a substitute for feed corn), would lead to a reduction in N runoff of around 3 per cent per hectare of grain and 11 per cent per hectare of grass in Norway (see Table 20.4).

In the scenario further liberalisation is used as a basis for free import of Danish grain, so that the maximum price for grain in this alternative would be the Danish wholesale price *plus* the cost of transport to Norway; this means a little more than half the price of the base alternative (feed grain prices are reduced from NOK 2.70 to NOK 1.16 and food grain falls from NOK 2.85 to 1.12). This would reduce N runoff in Norwegian grain production by more than 13 per cent per hectare (NOU 1991: 21B).

At a rough estimate, the rough feed price will fall by around 30 per cent. In this case it would give a reduction in N runoff per hectare of around 24 per cent. It is, however, important to emphasise that speculations about changes in rough feed prices are very uncertain, especially in the case of heavy reductions in grain prices. This is very obvious in Table 20.5, which summarises what the consequences of the two deregulation scenarios are for N runoff per hectare (given the above preconditions).

The reduction in fertiliser levels per hectare as a result of lower production will, in other words, lead to a significant reduction in N runoff

Table 20.5   Reduction in product prices and N runoff, per hectare, as a result of deregulation

	'GATT alternative' %	'Further liberalisation' %
Reduction in grain price	15–16	at least 50
Reduction in N runoff per hectare grain	3	13
Reduction in rough feed price	15	20–40
Reduction in N runoff per hectare grass	11	15–35

per hectare from agriculture. There is, however, a great deal of uncertainty about the size of these effects. The greatest uncertainty is linked to the runoff functions which the calculations in NOU 1991: 2B are based upon.

Reduced intensity would probably also contribute to a reduction in P runoff per hectare from agriculture, although these effects are more uncertain and would probably be far less than those presented for N fertiliser. This is because P pollution from agriculture is mainly a result of soil erosion, and the extent of this problem depends on the area of cultivated land which is in danger of erosion because it lacks plant cover during heavy rainfall in the autumn and/or in the snow melt period in the spring.

Reduced production of grain and animal products in Norway

To begin with, we assume that deregulation will not have appreciable consequences for the regional distribution of production and we will also ignore, for the time being, the effects of change in farm structures. When agricultural production and total agricultural land in Norway is reduced, total pollution levels from agriculture would, of course, be reduced. To what extent a reduction in Norwegian agricultural production from a wider pollution aspect would be positive is, however, a moot point and may vary from region to region. It also depends on the pollution problems we are interested in:

- the local (in Norway)
- the regional (North Sea)
- the global (Greenhouse effect).

Whether the flow of nitrate salts into the North Sea increases or decreases depends on from where Norway decides to import most of its food. If most of the food imports come from other North Sea countries (for example, Denmark or the Netherlands) the total impact for the North Sea would probably be negative. Much of present-day Norwegian animal farming takes place in western Norway and the Trøndelag region (with some also in northern Norway). No rivers from these regions drain into the North Sea.

It is also important to point out that relevant importing countries, including Denmark and especially the Netherlands, carry out far more intensive animal husbandry than is the average in Norwegian agriculture (see Lothe, 1995). Nor is there reason to believe that the gas losses from agriculture per unit produced in future import countries would be less than the average gas loss in Norwegian agriculture. The total emission of greenhouse gases would therefore probably increase as a result of an increase in total production of food.

The change to larger farm units

When agriculture is deregulated, profitability in concentrate-dependent production will be considerably weakened. Egg and meat prices will fall more than the cost reductions resulting from cheaper concentrates, so that the overall impact will be negative. In the calculations of the 'GATT' and further liberalisation alternatives we have allowed for a significant increase in the concession limits in concentrate-dependent production and greater milk production is 'allowed'. It is probable, practice, that the structural change, if deregulation goes far enough, would be far more extensive than the alternatives allowed for in the JORDMOD model.

We need also to consider briefly any pollution effects in the case of various types of structural change.

Larger grain-growing farms

Analysis of farm survey data shows that there is no reason to believe that consumption of polluting production factors per hectare (for example, chemical fertiliser and crop sprays) would increase in the case of a change to larger production units. The structural change may, however, lead to changes in farming techniques, in the sense of larger and heavier tractors and combine harvesters, as well as more special equipment for cultivation, fertilisation and sowing. Heavier machinery may have a negative impact on the environment, while an increase in expensive machinery and increased mechanisation in grain production could lead to significantly less soil erosion.

Larger livestock farms

Whether pollution problems in animal farming will increase or decrease is dependent on how much agricultural land per animal will be needed, compared to the present day. Analysis of farm survey data indicates generally that fertiliser costs per litre of milk produced fall in proportion to increases in herd size. In southern agricultural regions there were significantly lower fertiliser costs per litre of milk produced for larger herds than for smaller herds. This difference may possibly be a result of larger farms being overrepresented in areas of good production conditions for grass (the low-milk zone). So far, there is no basis for claiming that structural change in milk production will lead to either increase or decrease in pollution levels in animal farming, given that the average area of rough feed per head is not drastically changed.

All in all, there is reason to believe that changes in animal density, by which we mean changes in spread per head of the remaining animal farms resulting from any increase in agricultural specialisation, would have a far greater impact than any economies of scale from exploitation of environmentally sound technology for storage of animal waste.

Possible changes in regional distribution of production

Dairy production is by far the most important form of animal production with respect to both employment and land use. If a system of milk quotas is maintained, and the government does not allow quotas to be sold or transferred from one region to another, then the main characteristics in the present-day regional distribution of agricultural production will be maintained. If policy is changed to the free establishment of dairy farms and/or unrestricted sale of milk quotas, then milk production will probably increase in the best dairy areas – for example, Jæren. Nor is it improbable that dairy farming in the lowland areas of eastern Norway would increase in the long-term as a result of deregulation of Norwegian agriculture. In that case, dairy farming in the more marginal mountain and valley districts would simultaneously be reduced.

If profitability in concentrate-dependent production is significantly weakened, such production would probably be located in a restricted number of geographical areas where there is, or will be, a competent professional community, relatively low transport and concentrate production costs, and efficient abattoirs and a processing industry. It would seem likely that this would lead to an appreciable share of this production being localised in the Jæren region and/or the grain districts of eastern Norway.

The environmental impact of such a regional distribution of animal farms would be complex. If Jæren, relatively speaking, sees an increase in animal farming, there would be negative impact on local N pollution, because animal density is already high. As Rogaland, to which Jæren belongs, is not regarded as a North Sea county (in the North Sea Treaty) such an increase would probably have no negative impact on the North Sea.

An increase in animal production in the lowland areas will have a positive environmental effect, especially locally. Increased presence of animals eating rough feed would mean that much land in danger of erosion in eastern Norway would be used for rough feed production.

This would lead to reduction in soil erosion and a consequent reduction in P pollution from agriculture. If parts of animal production are moved from areas of high animal population per hectare (given that these drain into the North Sea), the transfer to animal-sparse areas (grain districts) gives a dividend with respect to the goal of reducing N runoff into the North Sea. If animal farming is moved from western Norway, the Trøndelag region and/or northern Norway then the result will be an increase in N runoff into the North Sea.

Possible changes in degree of specialisation in agriculture

At present, Norwegian dairy farmers produce most of the rough feed their animals need. Pork, egg and chicken farmers base their production for the most part on commercial concentrates. According to the law, animal farmers must farm an area of land which is large enough to satisfy minimum demands for spreading animal waste. Even on animal farms, in other words, there is a certain amount of arable farming, either production of fodder for the farm's own animals and/or production for sale of grain or other plant products. A significant share of grain production takes place on specialised grain farms.

Increased specialisation will lead to two things:

(a)   Parts of the agricultural land on many farms are at risk of erosion and are used on mixed farms as meadowland and pasture. Specialised grain farms would, however, use the land for crops, apart from the more marginal areas which would be uncultivated or planted as forest. The result of a farm changing from mixed farming to specialised grain farming is increased soil erosion.

(b)   An increase in animal density (reduction in the spread area per head) on remaining animal farms will lead to less of the N content of animal waste being used for plant fertiliser, while a larger part will disappear as runoff N and N gas loss.

In other words, there is reason to believe that increased specialisation will lead to greater pollution both of phosphates (as a result of increased erosion) and nitrates (increased N runoff and N gas losses).

On the other hand, a reduction in the degree of specialisation – as could take place if much of animal production was transferred to former specialised grain farms in eastern Norway and in the Trøndelag region – would have the opposite effect (Box 20.2).

In the JORDMOD model only specialised farms are modelled. There are certain aspects which speak both for and against agricultural production being increasingly specialised as a result of deregulation. Deregulation could mean:

- lower prices
- greater price variation (market-led prices)
- larger farms.

Box 20.2   Environmental gains from moving animal farming from Jæren to Romerike (eastern Norway)

A study has calculated the environmental impact of transferring animals from the densely animal populated Jæren to Romerike. In one of the alternative calculations the consequences of moving much of the 'rough-fed' animal population from Jæren to Romerike was examined. It was assumed that the animals were distributed among the existing farms. As a result of the movement the population density in Jæren sank and the Romerike farms became less specialised.

In this alternative it was assumed that only some of the animals were moved to Romerike. The positive environmental impact was mainly a result of 'the improved exploitation of animal waste which became possible after restructuring'. This improvement in exploitation of animal waste can be seen through a significant reduction in consumption of nitrous chemical fertiliser. In addition to the reduction in runoff nitrates, soil erosion was also reduced in Romerike by the change, as the introduction of rough-fed animals will have also led to an increase in production of rough feed and a reduction in grain production in areas at risk from erosion. The model was not capable, however, of quantifying the impact.

Source: Holm and Holm (1989).

When agricultural prices fall, farms have to produce with lower capital costs per produced unit, or marginal income will cease. The consequence of deregulation may be an increased degree of specialisation in agriculture. There are, however, a number of conditions which will, in any case, moderate this tendency towards specialisation:

- The majority of present-day grain farmers are part-time or hobby farmers. A change to significantly larger and fewer farms will mean that a greater proportion of the arable land is cultivated on a full-time basis. Some of these full-time users may find it profitable to produce other crops than grain in order to exploit their machinery and labour better in periods of spare capacity.

- Market-led prices lead to increased risk in agriculture. It is probable that spreading risk, by investing in more than one type of production, would become more important if farmers did not receive politically guaranteed grain prices. It is also probable that it will become far more important for full-time farmers (who depend on the income from their farm) than for the part-time farmers (who have another income as security) to spread risk by investing in several forms of production.

In other words, it is not possible to come to any unambiguous conclusion about whether the tendency towards specialisation will be reinforced or weakened as a result of deregulation.

20.7   The impact of further deregulation on biodiversity

Approach and problem clarification

The agricultural countryside can be roughly divided into home fields and residual biotopes. The home fields can vary widely, depending on production, farm type and natural conditions. We often regard the home fields as the dominating element, while the residual biotopes are 'islands' in the stream. This is the situation in the very best Norwegian agricultural areas, where the home fields can make up half of the land area in a municipality. On the other hand, in the more marginal areas it is natural to regard arable land as the 'islands'.

It is important to clarify the structure of the Norwegian countryside and interpret this pattern if we are to be able to understand the ecological processes which affect the system. The explanatory models for these processes can form the basis of assessment of the significance of changes in field sizes and shape, production specialisation, treatment of residual biotopes and fallowing of the home fields.

A cross–hierarchic division of ecosystems into composition, structure and function can contribute to a better understanding of diversity in the system (Noss, 1990):

(a)   Composition is linked to the number of habitat types, species, populations and genes.

(b)   Structure is linked to heterogeneity in the landscape, on various scales.

(c)   Function is linked to ecological processes in time and space
      (Franklin, 1993; Cale and Hobbs, 1993).

So far, diversity linked to composition has been overfocused, while
structure and function have not been widely discussed.

A hierarchic model assumes that higher levels of organisation incorporate, limit and regulate the process at lower levels. This indicates
that structure at a relatively rough level can contribute to increased
understanding of the dynamics, giving the authorities a rough tool to
work with. The model calculations which we have referred to are aggregated at a high level. It is, therefore, also natural to adapt to this level
in the discussion of biodiversity, but we shall focus on a lower scale
level where it is natural to do so. Based on the model calculations
above, there are three trends in a further deregulation of agriculture
which will have an indirect significance on biodiversity:

1. lower product prices
2. reduced production
3. increased average sizes of production units.

We shall take a closer look at how the following types of change in
production will affect the biodiversity of species at a relatively rough
scale level, based on the possible consequences on production of the
changes in these framework conditions:

(a) reduced production intensity
(b) reduced land under cultivation
(c) continued structural rationalisation
(d) changed regional distribution
(e) changed degree of specialisation.

In the following discussion we will not address the traditional agricultural land types. These form an insignificant proportion of agricultural
production and the model calculations above tell us nothing about the
consequences for such farms.

Reduced production intensity

Agriculture means a manipulation of natural systems, creating an agricultural countryside. This is the basis for efficient production of food.
Use of man-made production factors has increased heavily over the last
50 years. In the areas of intensive farming, reduced profitability can

lead to a number of marginal areas becoming unprofitable to farm, and they will be laid fallow. This may be positive in an interim phase, especially if the new growth takes place 'naturally' without replanting (Ekstam, Aronsson and Forshed, 1988). The liberated arable land may become important biotopes for flora and fauna, as long as efficient management takes place. Ensuring a good link between the various landscape elements can contribute to joining various sub-populations together, strengthening their chances of survival (see, for example, Mader, 1988; Taylor *et al.*, 1993; Berg and Pärt, 1994).

If the authorities maintain relatively high land and animal subsidies, low product prices may lead to a greater part of the agricultural land being used for extensive grass production. More extensive animal husbandry may lead to greater use of partially natural grazing land, especially for calves and beef cattle. These pastures are often species-rich, in plants, insects and not least groundsoil fauna (see, for example, Steen, 1980; Matthey, Zettel and Bieni, 1990; Gibson *et al.*, 1994). They also often function as transit zones between intensively farmed land and the original nature. These diffuse transit zones are important concealment biotopes and escape routes, as well as being important wintering areas for insects.

Extensification of farming will mean that much marginal agricultural land would be treated with chemical fertiliser. This will mean better conditions for many stress-tolerant plant species which die when exposed to nitrogen (see, for example, Glimskär and Svensson, 1990; and Kielland-Lund, 1991). The presence of flowering plants, which may be very common in unfertilised meadows and pastures, will be decisive for many nectar and pollen-eating insects (see, for example, Mikkola, 1987). Lower product prices will also make it profitable to reduce the use of crop sprays. Use of sprays is relatively low for climatic reasons, and has been reduced over the last few years as a result of extensive guidance programmes and introduction of programmes for integrated disease and insect control. Increases in knowledge of biological control in the countryside will contribute to a further reduction. Spraying is harmful for biodiversity in the border zones and in the meadows. Most meadow and pasture insects depend on weeds as food (Hanski *et al.*, 1988). Any reduction in spraying is, therefore, important both for the weed flora and the insect fauna (see, for example, Fogelfors and Bjørkhem, 1986; Esbjerg 1987; Dover, 1991).

Reduced cultivation as a result of reduced intensity would probably lead to a richer groundsoil fauna. Many surveys show increased diversity in the groundsoil fauna when ploughing is reduced (see, for example, Matthey, Zettel and Bieni, 1990; Stinner and House, 1990).

The general land and agricultural (LA) subsidies form a significant part of production subsidies in agriculture. These subsidies contribute to keeping marginal arable land under cultivation. There are also a number of general demands linked to the subsidies, including bans on spraying borders and copses. These demands can be extended and contribute to securing biodiversity. This requires, however, supervision and control on the part of the authorities. One unfortunate aspect of the LA subsidies, as they are drawn up, is that formerly marginal areas, unfertilised meadows and pastures, are fertilised to increase farms' arable land and therefore subsidies (DN, 1994). The Ministry of Agriculture is in the process of revising the regulations to prevent such undesired effects.

Reduction in land under cultivation

According to the calculations we have made above, there is reason to expect a certain reduction in agricultural land, both in the south and north. Much marginal agricultural land has ceased to be cultivated over the last fifty years. However, the total area of agricultural land has remained relatively stable over the same period. This means that relatively large areas have newly come under the till, especially in the better arable districts. These trends have been reinforced over the last decade, especially in northern and western Norway, and in Agder. In parts of northern Norway and inner Agder half of the home fields are kept fallow. This probably has major consequences for the degree of isolation for many populations of farmland species and therefore represents a risk of loss of diversity both locally and nationally.

In Norway's best grain districts around 50 per cent of the land is under cultivation. A certain reduction in the agricultural area in these districts may have a positive impact on diversity. The interesting point here is the alternative use of superfluous land. Forestry and extensive grazing are two relevant farming methods (Moen and Klynderud, 1994). Planting with spruce would not be viable, based on the wish for high biodiversity (Ekstam, Aronsson and Forshed 1988). Frivold (1993) discusses the choice of tree species in reforestation in the mixed farming situation, and concludes that dead wood and a high proportion of deciduous trees are important for diversity.

Any surplus of agricultural land would contribute to weakening soil protection, especially in the exposed areas. This would lead to increased reduction of agricultural land and therefore depreciate the total diversity.

Relatively large reforestation programmes are being carried out in Western Europe to increase the share of 'archetypal' nature in agricultural countryside and to ensure that these link up to a greater extent (Merriam, 1988; Saunders, Hobb and Erlich, 1993; and Rønningen 1994b). This is not particularly relevant in Norway, as there is little to indicate that the fragmentation of the forests contributes to reduction in the extent of birds and mammals (see, for example, Fry and Main, 1993; Andrén, 1994). It is accepted that as long as 30 per cent of the original biotope is intact, the spread of birds and mammals is not in danger (Andrén, 1994). Little is, however, known about plants' and insects' ability to spread on the small scale. For such species, the most important thing is to ensure a small-scale mosaic. Such restoration measures would, however, be very significant if lifeforms are to thrive in the agricultural landscape.

Agriculture has laid claim to the most productive types of nature. On the local scale, restoration of such rich types as wetlands could contribute to increasing the diversity of types of nature and therefore the diversity in agricultural countryside. Low profitability in agriculture will also limit further cultivation of such valuable types of nature and may therefore have a positive impact on local diversity.

Continued structural rationalisation

Norway's natural conditions are a hindrance to the development of large scale agriculture. Together with farming and professional combinations, this has created the basis for the maintenance of a small-scale agriculture. The general economic and technological trends have, however, led to the number of farms being halved over the last 30 years. Larger farm units and increased mechanisation have contributed to larger fields with reductions of borders and removal of corridors (gullies, ditches, field and property borders). It is, however, difficult to quantify the extent of this. Even if the farm units have become larger, the ownership structure has not changed to any degree. The ownership structure is important for the physical structure of agricultural countryside. If this breaks down, many linear elements which hold the landscape together will disappear; this applies especially in the lowland districts where the topographic conditions allow larger continuous fields. It may possibly be balanced up by official requirements for the maintenance of such elements as one of the concession conditions when acquiring property. In marginal areas the topography will limit size of fields and therefore the impact of an increase in farm units will be less.

The labour per hectare for the production of grain and grass levels out in the case of large fields. The gain from increased field size is less for grass than for grain (Christoffersen, 1995). A rough estimate indicates a relatively small gain from fields which are larger than 40 hectares for grain and 20 hectares for grass; most Norwegian arable fields are smaller than this. Calculations also show that the shape of the fields has a lot to do with the labour efforts per hectare. Long, straight fields need the least labour per produced unit. This indicates that there is relatively much to gain from a better shaping and extension of fields in Norwegian grain and grass production. Such efficiency measures can also deteriorate local and partly regional diversity, especially in lowland areas where conditions are suitable for such change.

Changes in regional distribution

We have seen a strong regional displacement of production since 1950. This is mainly a result of deliberate canalisation policies by the government (Gabrielsen and Vatn, 1988; Gillebo, 1990) as part of regional policy. The lowland districts have tripled their area under grain, while there has been an equivalent reduction in the mountain and valley districts. Dairy production has to a great degree vanished from the lowland districts and moved to the mountain and valley districts.

This regional displacement of production has had especially negative consequences for biodiversity in the lowland districts, which today are only under grain. Even if production in the mountain and valley districts is just as one-sidedly directed on dairy production, the properties are smaller and not as concentrated. There is extensive rough-feed production which gives a more varied landscape than with grain production.

For the sake of the environment, it would be desirable to return part of the dairy production to the lowlands so that the rough-feed area forms a minimum of 10–20 per cent. This could form the basis of more universal farming at the district level and therefore make a positive contribution to biodiversity.

Changes in degree of specialisation

General economic and technological developments have led to an increased degree of specialisation in the agricultural sector. Farmers have eliminated variation within the individual field through a mono-culture of crop plants and through equal cultivation and fertilisation.

At the same time, they have created a new form of variation between fields with different crops. This has formed the basis for species which tolerate a large degree of disturbance. A further specialisation of production will, however, reduce this mosaic and therefore have a negative impact on diversity at the farm level. The diversity of produce at the farm level has declined drastically during the last 50 years. The reduction in the number of crops (number of agricultural plant types) is regarded as an important negative factor for local diversity of species. This leads to a less varied landscape.

The changes in the physical structure of the landscape has been well documented in Rygge municipality (Moeini, 1988) where a decrease of about 20 per cent in the number of corridors was found. It seems likely that this change is largely a result of increased specialisation in agriculture.

Most modern Norwegian farms are specialised. It is uncertain whether deregulation would lead to further specialisation, or whether the change to larger farm units would increase the number of mixed farms (increased competition can be seen as a motivating factor for greater risk-aversion and therefore more varied farming methods). Mixed farming methods at the farm or district level would contribute to better agronomics, and therefore increase diversity. As an example, animal dung has a positive effect on the structure of the soil and the content of organic material and increases the diversity of species in the field (see, for example, Pimentel and Warneke, 1989; Matthey, Zettel and Bieni, 1990).

Agriculture and biodiversity: a further perspective

The goal and the motives for protection and management of biodiversity are often unclear – concentrate on 'freezing' present-day biodiversity, restore the agricultural countryside to a given era, or create anything qualitatively new to increase diversity?

Sustainable agricultural methods are a precondition for a rich diversity which is, in turn, necessary for a sustainable agronomy. Securing a functional agriculture will, therefore, form the key for a rich flora and fauna in the countryside.

Changes in the diversity of species as a result of agricultural change may be large in the individual farm, while at a higher level the local variation may be erased by different farms choosing different farming methods. The management of biodiversity must operate on several scales, and this is further complicated as species also operate on several scales.

Food production involves manipulation of the environment and must necessarily come into conflict with preservation of diversity. Organisation and distribution of production has many social implications which indirectly affect diversity. In a production system it cannot be taken for granted that measures which are positive for local diversity will be positive on the regional level. The same applies nationally and globally. A transfer of parts of Norwegian animal production to the Netherlands may contribute to preserving diversity in Norway, but will lead to further deterioration of the environment in the Netherlands.

21
Possible Consequences for the Transport Sector of Increased CO_2 Taxation*

21.1 Cost trends under various CO_2 taxation schemes

The Project for a Sustainable Economy has developed a model to study the tax load on various forms of transport if a CO_2 tax is levied. The study is mainly based on figures from a report from Vestlandsforskning (Heiberg and Høyer, 1993), studying the following forms of transport:

– private cars
– express coaches
– diesel trains
– aircraft (Twin Otter from Widerøes Airline)
– high-speed ferry.

In Norway, in 1985 there was a demand for 41 477 million person km of transport using motorised means of transport, equivalent to around 10 000 km per Norwegian. As we can see from Figure below, travel by private car formed around 80 per cent of this amount. For every 10 000 kilometres travelled by each inhabitant each year, 8000 were by private vehicle. For this reason we have placed the main emphasis of our analysis on travel by private car.

The analysis looks only at emissions resulting from direct energy use. This means that emissions connected to the manufacture of cars, construction and maintenance of infrastructure, or production of the energy consumed by the means of transport, are not considered. The analysis is thus connected only to consumption of fossil energy in the transport process and the emissions resulting from it. The time

---

* This chapter is built on Jespersen (1995).

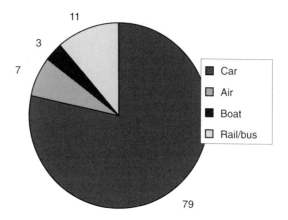

Figure 21.1 Total number of person km 1985, distributed by mode of transport (per cent)

frame is the same as in the long-term scenarios of the MSG model – that is, from 1989 to 2030. Our basic premise is that cars, buses, trains and aeroplanes will all use the same fossil fuels that they use today until 2030. We know that this is hardly likely, and there will probably be a movement towards increased use of natural gas, methanol, hydrogen and electricity. It is our opinion that the resource energy efficiency improvement that we meet over this period of time will be so large that it will encompass the change to more resource-friendly technology. First, we shall examine the impact on cost trends for motor transport as a result of introducing the alternative policy regime we have studied above by means of the MSG model.

Travel cost trends for private cars

The analysis builds on the following central assumptions (Heiberg and Høyer, 1993):

– The average consumption of fuel per vehicle/kilometre in the first year is 0.098 litres of petrol.
– Capacity utilisation is 1.75 persons per vehicle, remaining constant over the whole period.
– Combustion of 1 litre petrol emits 2302 g of CO_2.

In the following analysis we study cost trends for passenger transport in three of the alternative scenarios (as discussed in Chapters 16–18). These are:

Alternative 1 The significance of reducing oil and gas extraction in especially vulnerable areas (in this scenario, CO_2 taxation is the same as in the base alternative in the Government LTP).

Alternative 2 The significance of high CO_2 taxation.

Alternative 3 The impact of combining high CO_2 taxation with reduced oil and gas extraction.

We have also chosen to carry out a sensitivity analysis of the significance of various technological achievements for fuel consumption and CO_2 emission, and through them the travel costs for the different means of transport. We assumed three different trends:

– The 'zero' trend means that there is no technological achievement.
– The 'moderate' trend means that we assume a reduction in the specific fuel consumption of 0.7 per cent per person km per annum as a result of improved technology.
– The 'optimistic' trend means that we assume a reduction in the specific fuel consumption of 1.5 per cent per person km per annum as a result of improved technology.

As far as improvements in energy efficiency for the private vehicle fleet are concerned, we regard the potential for change to better and newer energy to be sufficiently large for us to concentrate on the zero development and the optimistic development trends when illuminating the significance of technology on travel costs.

First of all, we shall examine how petrol prices will develop as a result of the policies in the three alternative scenarios. Figure 21.2 shows the trends in 1989 NOK for alternatives 1–3. We can see that in scenario 2, the real price of petrol virtually doubles until 2010. The reason for the price of petrol peaking in 2010 (at NOK 12.42 per litre) in this scenario is that the real value of the CO_2 tax also peaks at this time. It does, however, fall from NOK 12.42 in 2010 to NOK 9.98 in 2030. In alternative 3, it reaches its peak in 2023, as this is when the CO_2 tax reaches its highest real value in this alternative. There is a completely different situation in the case of scenario 1. The real value of the 1989 tax (NOK 320 per tonne of CO_2) falls, which means that fuel becomes cheaper every year. The real price of fuel (NOK 1989) falls to NOK 6.09 in 2030.

Will a petrol price of more than NOK 12 per litre (NOK 1989) place a stranglehold on private travel around the country? We must point out again that private cars are becoming more and more energy-efficient, which will mean a considerable saving in fuel consumption. Figure 21.3

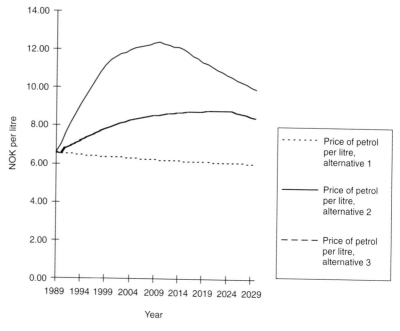

Figure 21.2   Petrol price trends, per litre, for three alternative scenarios, 1989–2030

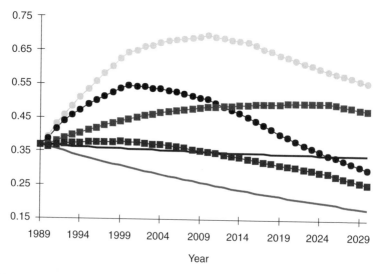

Figure 21.3   Total fuel cost trends (NOK) per person km, for differing CO_2 tax and technology trends for private cars, 1989-2030 (NOK 1989).

shows the real cost per person km to car travellers using different technologies. This shows how significant technological advances are for the costs.

The various alternatives are marked in Figure 21.3 by different symbols. The alternative 1 curve is a line, the alternative 2 curve uses circles and alternative 3 uses squares. The highest curve for each alternative represents a fuel-efficiency growth of zero, while the lower curve represents a 1.5 per cent fuel-efficiency improvement per annum. As we can see from Figure 21.3, the zero-technology development in alternative 1 would be yield a slight reduction in travel costs throughout the period. This reduction is the result of an initial reduction in the real value of the CO_2 tax and means that it would be 7 per cent cheaper to travel in 2030 that in 1989. If we consider the optimist technology trend in alternative 1, we can see that this leads to a drastic reduction in costs throughout the period. According to this scenario, it would be more than 50 per cent cheaper to travel by private car in 2030 compared with 1989. This is a result of both reduced price of petrol and reduced costs of the CO_2 tax following reduced emissions, due to reduced specific fuel consumption due to technological progress.

In alternative 2, the zero-technology trend leads to an increase in travel costs which peak at 88 per cent above the 1989 level in 2010. After this, travel costs show an annual fall. According to this scenario, motoring will be 48 per cent more expensive in 2030 than in 1989. If we look at the optimist technology trend in this alternative, we observe that travel costs peak as early as 2000. At this point in time, travel costs will be 48 per cent higher than in 1989, although at the same time 40 per cent lower than the zero trend for this alternative. After 2000, travel costs will show a rapid decline as a result of reduced taxation and fuel costs following from technological progress. In 2022 they will have fallen to the 1989 level, and in 2030 they will be 17 per cent lower than in 1989.

This shows the significance of fuel economy for motoring costs. We shall now examine how travel costs could be thought to change for other means of transport in the analysis when compared against probable technological advances.

Travel cost trends for public transport

In the following we shall examine four different means of transport – that is, express coach, diesel train, air and high-speed ferry, illustrated in alternative 2. For a more thorough examination of these means of transport and the premises on which they are based, see Jespersen (1995).

As far as public transport is concerned, we shall assume that these forms of transport do not have the same fuel-economy potential as the private car. The reason for this is that the rate of renewal in the fleets is lower, competition between car manufacturers is harder than manufacturers of these vehicles and the demands on private car fleets outside Norway are stricter than for public transport (for example, air transport is exempted from CO_2 taxation). These means of transport are not manufactured in Norway, so that it is unlikely that a national Norwegian CO_2 tax, as described in this analysis, will motivate manufacturers to speed up technological advance. We assume therefore that development will take place at moderate levels. High-speed ferries are the exception to this rule. Norway is in the forefront of development of this form of transport, making it reasonable to assume that a high level of CO_2 taxation will force the pace of technological development. We would therefore expect the optimist trend to apply for high-speed vessels.

It is important to be aware that travel cost for public transport passengers is a synonym for ticket price. Contrary to car transport, these costs do not just contain fuel costs. The relative changes in fuel costs will, therefore, form a smaller proportion of the travel costs than for private cars. This moderates the conclusions of the analysis. This aspect has been discussed in more detail in Jespersen (1995).

Figure 21.4 shows possible travel cost trends for the various means of transport when taxation is phased in according to alternative 2 and with the expectations to technological advance discussed above.

First, let us examine express coaches. It is worth noting here that express coaches have fundamentally the lowest fuel consumption of all of the transport in this analysis. The result is that change in travel costs remains fairly moderate. Express coaches achieve their maximum cost level in 2006, when it will be almost 9 per cent more expensive to travel by coach than it was in 1989, measured in 1989 terms. After this costs fall annually, so that by 2030 it will only be 3.6 per cent dearer to travel by coach than it was in 1989.

Diesel trains have a somewhat higher specific fuel consumption than coaches. This means that the taxation will hit trains harder. Travel costs peak in 2006 and end at 2030; 8.3 per cent higher than in 1989.

One of the characteristics of air travel is its very high specific fuel consumption. Fuel costs are, however, only a small part of ticket prices. This means that changes in travel costs will be relatively modest. These peak in 2006, when they are around 20 per cent higher than in 1989.

As far as high-speed ferries are concerned, as we have pointed out above, we expect more rapid technological advance for these means of transport as a result of increased Norwegian CO_2 taxes than for the others. It is worth noting that high-speed ferries, according to our

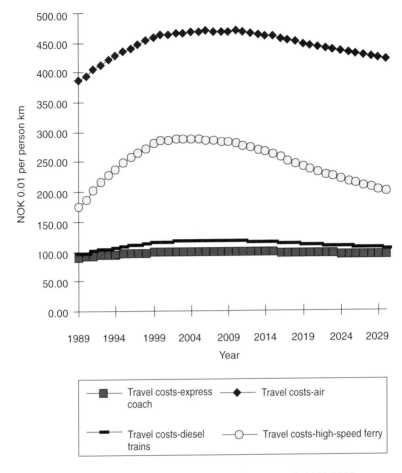

Figure 21.4   Travel costs for public transport, alternative 2, 1989-2030

figures (Heiberg and Høyer, 1993), are indisputably that means of transport which uses the most fuel per person km. Fuel costs also form most of the ticket price. These vessels will be dearest in 2002, when travel costs will be 64 per cent higher than in 1989. As fuel consumption is such a large part of the ticket price, the potential for a reduction as a result of more rapid technological advances is very great. This means that travel costs are reduced drastically over time, and in 2030 it will only be 14 per cent more expensive to travel by high-speed ferry than it was in 1989, measured in 1989 prices.

We sum up some of the relevant figures in Table 21.1.

Table 21.1 Relevant figures from the analysis, 1989 and 2030 NOK, 1989 prices

Means of transport	Fuel consumption per vehicle/10 km 1989 (litre)	Fuel consumption per vehicle/10 km 2030 (litre)	Emission of CO_2 per vehicle/10 km 1989 (gram)	Emission of CO_2 per vehicle/10 km 2030 (gram)	Fuel price per litre maximum (NOK) (2010)	Fuel price per litre 2030 (NOK)
Private car	0.98	0.525	2 257	1 207	12.83	10.01
Express coach	3.03	2.42	8 629	6 328	9.29	6.67
Diesel train	39.8	30.0	104 077	78 450	9.19	6.57
Air	16.1	12.1	40 056	30 104	8.52	6.03
High-speed ferry	137	102.8	366 612	275 092	8.99	6.32

In the next section we shall examine in more detail how these changes in travel costs affect the demand for various means of transport.

21.2 The impact of travel costs on travel demand

The demand for person km of transport by the means we have discussed is influenced by various factors, including the change in relative travel costs. One model from the Norwegian Institute of Transport Economics (Ramjerdi and Rand, 1992) is used to examine to what extent demand changes when the same taxation is levied on all of the means of transport. We illustrate this using alternative 2, partly because it has the strongest phase-in CO$_2$ tax, and partly because Statistics Norway's partial survey (Alfsen, Larsen and Vennemo, 1995) concludes that it is the most cost-effective method of reducing CO$_2$ emissions. As above, we shall use the optimist trend for technological advances in the car and high-speed ferry fleets. In the case of the remaining modes of public transport we shall, as we have previously described, use the moderate trend. Figure 21.5 shows how the trends in travel costs which we have described will affect demand for person km.

We can see from Figure 21.5 that the transport which will be hardest hit by the CO$_2$ taxation is the high-speed ferry, while the express coach is that which is least affected. It is also worth noting the trend for demand for private cars. We note that it shows a drastic fall of about 7 per cent before rising until 2030, when it ends at nearly 4 per cent above the 1989 level. This is partly because petrol costs make up all of the variable costs, giving taxation a fairly heavy impact on demand. This explains the initial rapid fall. The rise in demand again can partly be explained by technological advances. This means that fuel consumption falls every year, which first of all reduces fuel costs. Reduced fuel consumption also reduces emissions. This will particularly begin to apply after 2015 when the taxation in-phasing stops; it is also index-linked and this effect also begins to show. An average inflation of nearly 3 per cent leads to a major reduction in the real value of the tax. As regards the total change in demand for express coaches and diesel trains, the change is somewhat larger than is realistic because the Transport Institute's model does not contain any cross-elasticity between the two means of transport.

21.3 What the sector analysis cannot answer

It is important to be quite clear that our discussion has not considered the total composition of demand. We have concentrated only on

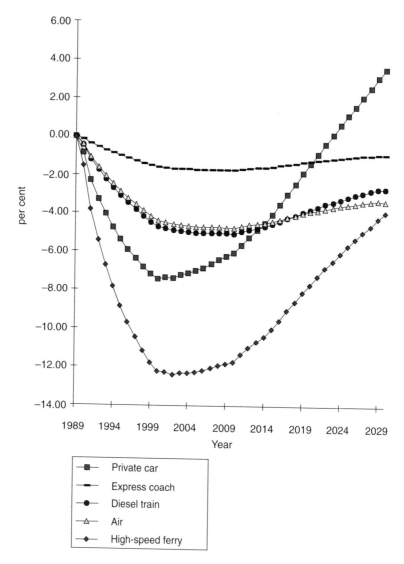

Figure 21.5   Demand for person km, 1989–2030

finding what the effect would be on demand, in isolation, when we examine three different technological trends. This means that any increase in transport demand as a result of an increase in real income would not be disclosed by this analysis. It is also important to be quite aware that the presumption that all cost increases resulting from an increase in taxation will be transferred to passengers is uncertain. On the one hand, it may be that any transport which is in a situation of sharpened competition would not transfer the entire cost to the passengers. On the other hand, it may be that a means of transport which lacks competition will take the opportunity to overcompensate – that is, to increase ticket prices more than purely required by the tax increase.

Electric trains are not included in the analysis because fossil fuels are not used. In our model emissions will be virtually zero, so that no direct effect of a CO$_2$ tax will be present. There is, however, reason to believe that electric trains would benefit considerably from price increases in the other means of transport. The data is based on full-price tickets and ignores all discounts and cheap rates. This means that the ticket price for the various means of public transport could be somewhat higher than the average ticket price, and the effects on demand somewhat larger than this analysis shows.

In addition to the limitations discussed above, this sector analysis exhibits the following weaknesses:

- The analysis includes only the effects on transport for journeys of more than 100 km. This means that we cannot see the effects on local urban and suburban transport.
- The basic data for air transport is based on Widerøe Airline's Twin-Otter aircraft. The weakness of not including Braathen SAFE and SAS is obvious; these airlines are not included because the data is not available.
- The analysis is based on three main sources (Alfsen, Larsen and Vennemo, 1995; Heiberg and Høyer, 1993; Ramjerdi and Rand, 1992). The analysis will exhibit the same weaknesses as those inherent in these sources.
- We have made basic assumptions about technological advances in fuel consumption; there is a very high level of uncertainty around these.
- When national taxation is introduced, trade leakage takes place. This has not been discussed in this study.

All the above points are ambiguous. It is, therefore, very difficult to make any definite statement about whether they are positive or negative effects.

21.4   Conclusion

The analysis does, however, provide us with a great deal of useful information which we can divide into three main points:

1. If the CO_2 taxation does not increase (this actually means that it decreases in real value terms), those forms of transport which use most fuel will experience the most favourable demand trends, given that all other conditions remain constant. There are two reasons for this. First, it will be these means of transport which, in the year of introduction, pay the most CO_2 tax. This tax burden will be reduced, partly as a result of the consumer price index (CPI), and partly as a result of technological advance. In the second place, this transport will have the greatest potential for reducing fuel consumption. This means that advances in technology will make it relatively cheaper for them over time, because their fuel costs will fall more.

2. The tax burden will seem heavier on those forms of transport where fuel costs make up the most of the ticket price. This means that private cars and high-speed ferries will feel the increased CO_2 tax the greatest, while air transport will feel it least.

3. If we ignore the demand effect for the ferries, the conclusion of the analysis will be that these conditions are not dramatic for the transport sector, especially as all of the means of transport will experience a positive income effect which is not covered by this partial sector analysis. In practice, it means that there will be an increase in the transport sector for all the means of transport during the period, even if the taxation, taken in isolation, has a negative effect. At the aggregate level this income effect has been incorporated in the various MSG simulations with alternative CO_2 tax regimes in Chapters 16–18.

22
More Efficient Taxation from Environmental Policy[*]

22.1 Introduction

In Norway, as in the rest of the Western world, there is an ongoing debate as to what extent it is possible to change tax structure from taxation on labour to taxation on environmental grounds. This was the subject of a theoretical discussion in Chapter 12. However, one decisive question is the size of the revenues which it is possible to collect from increased environmental taxation, and which would enable a reduction in employers' contributions. At present, employers' contributions provide revenues of around NOK 50 000 million for the exchequer. We shall now examine the potential gains from an environmental tax, different forms of taxation and what obstructions we are likely to meet in the economy.

Throughout this book we have identified important negative environmental impacts of human activity. These may be local, regional or global. We have also discussed the opportunities for internalising such effects by applying various measures, which include environmental taxation on emissions or on use of polluting production factors. We have calculated the impact on both economic and environmental indicators of, for example, the introduction of Norwegian CO_2 taxation at levels above that envisaged by the government in its Long-term Programme (LTP).

First of all, we shall examine a little more closely whether the budgetary results of the MSG computations in our scenarios allow any significant system of fiscal change from tax on labour to tax on emissions. Then we shall look at the results from the newest computations carried out on behalf of the Project for a Sustainable Economy using

---

[*] This chapter is based mainly on Mathiesen (1995) and Hervik (1995).

the MISMOD, in which both macro- and sector-effects are focused on (see Chapter 19). Finally, we shall compare these with the results from an analysis of equivalent opportunities and limitations in the major EU countries, in which a number of environmental indicators have been attempted to be internalised by taxation.

Apart from this, we shall focus on another area which may be interesting from the point of view of a national and local reallocation of taxation and resources, and therefore from the environmental perspective. In this case, we are thinking of the so-called 'extra return' from scarce resources – in other words, the return which, for various reasons (to which we shall return to in some detail) occurs, and which occurs in addition to normal income and return on labour and capital. Apart from an overall environmental perspective, the main argument for taxing this extra return is that it satisfies the demand for efficiency, because we will not experience any allocation loss when we levy it. If the tax on the additional return is levied on behalf of the community it may provide the basis for reduction of taxation on labour, since allocation losses are assumed to be high under the present Norwegian fiscal system and taxation levels.

Now let us look briefly at some of Norway's most important raw material industries – petroleum, hydropower and fishing – in order to illuminate the extent of the potential and actual extra return. In addition, we shall look at ground rent in both urban and rural districts, and how it is affected by public infrastructure measures. Then we can show what potential there is for taxing these industries and what institutional barriers affect the collection of the extra return.

22.2   Will environmental taxation permit reduction of labour tax?

Long-term projections

The MSG model computations of the various future alternatives under CO_2 taxation show different paths of development for the balance of public budgets. Alternative 5, which provided increased leisure (Chapter 19), seemed to lead to a difficult budgetary situation. In alternative 2, where we unilaterally levied a high, steadily increasing CO_2 taxation over the next 25 years, we initially experienced an increasing and fairly substantial budgetary surplus, followed by reduction in the surplus, then a deficit which was on a par with the reference path in 2030. The long-term computations thus show that there is room for reduction in other taxes as a result of CO_2 taxation in the short and medium term, but in the long term the surplus is eroded by increasing budgetary deficits. This indicates that care must be taken not to regard the revenues from Norwegian CO_2 taxation as a permanent replacement for, (for example), employers' contributions, or as a basis for reduced levels of income tax. It is, however, important to remember that all of this reasoning is built upon the assumptions of the selected model. The possible trends that we have described are not presented as a description of what will happen, but rather as an illustration of a possible problem with the sort of environmental taxation in an economy which works as described in this macro-model.

Effects of revenues from a CO_2 tax on the public budgetary balance

Using the SNF model which was presented in Chapter 19 to study the effects on employment of the changes in CO_2 taxation, we can examine the impact on welfare of increased public revenues and the possibility of using the CO_2 tax in the short and medium term (until 2000) to reduce levels of other taxes – for example, employers' contributions. If possible, we want to bring out all of the consequences of changed general conditions in the period which is being examined. Real and financial investment does, however, affect the production and consumption in later periods than those where they take place. If two alternative scenarios mean different investment they would, in addition to different consequences within the period of analysis, lead to different consequences in one or more subsequent periods (after 2000), which would complicate a comparison of the scenarios. Just as importantly, we do not have, in any case, a good theory for investment within the framework of a static model: investment decisions mean comparison of payment flow over many periods and are best treated in dynamic models. For this reason we keep real and financial investment (measured in fixed prices) constant and equal over the various scenarios.

The budgetary balanced is assumed in the model as applying to the economy as a whole. Gross investment and the balance of trade are decided outside the model by the premise providers, while public and private consumption are decided by the model.

The public sector receives funding from capital, taxation and other charges. It purchases products and services (public sector consumption), saves and makes transfers to households (grants and pensions) and companies (grants and subsidies). This may be expressed as follows:

$$T = p_{Cpub} \cdot C_{pub} + p_{Jpub} \cdot J_{pub} + Z$$

Where T is net tax revenue, C_{pub} and J_{pub} are public sector consumption and investment, respectively, measured in fixed prices and interpreted as volumes, while p_{Cpub} and P_{Jpub} are the corresponding prices and Z is payments to households. As we have already said, investment is exogenous. Public sector consumption is not a central theme of the analysis, and this is why we have stipulated C_{pub} and maintained it at the same level in all scenarios. (In analyses of environmental harm from NO_x, SO_2 and the like, we have included possible savings in C_{pub} to allow for reduced emissions.) Because T is endogenous, it is dependent on prices and volumes across the whole economy, and because the prices p_{Cpub} and p_{Jpub} are endogenous, the demand for a balanced public economy will mean that Z must be endogenous. In brief, all that remains of the total fiscal revenues after public sector consumption and investment are returned to private households. In a GNP connection, it is therefore natural to use private sector consumption C_p (measured in fixed prices) as an indicator as to whether a solution means an improvement or a deterioration in the economy. In addition, we take into account the demand for leisure, so that the welfare goals in the analysis are an aggregate of private (goods) consumption and leisure.

A reduction in Norwegian emissions of greenhouse gases will mean a restriction on Norwegian freedom of action. It is therefore tempting to conclude that this will lead to a cost in the form of lost welfare. In a number of analyses there is also a cost at an emission reduction of around 15–20 per cent which is estimated to be in the range of 1 per cent of GNP (see NOU, 1992:3, KLØKT computations; Mathiesen, 1991; Nordhaus, 1991; Manne and Richels, 1994, indicate an average cost of 1.4 per cent of GNP, in a range from 0.3 to 7.2 per cent.)

As we showed in Chapter 19, this hardly need be the case. Use of environmental taxation can actually be thought to increase welfare. The central point is that in the theoretical, and in many numerical, analyses it has been explicitly or implicitly accepted that the economy is initially optimally adapted. In economic analyses of this question, present taxation and other deviations from a free trade equilibrium – for example, pollution – are often ignored. In such cases important characteristics of the economy are poorly represented (see the analysis in Chapter 12). These deficits do not necessarily have to be significant for the results. Because a tax-aided CO_2 reduction provides greatly increased tax revenues, while at the same time reducing emissions of other gases and road traffic, it has proved that these omissions are significant. We know that a further restriction must cost something. We find, however, the basic presumption to be implausible, and believe that few, if any, economies can be regarded as being in the vicinity of optimum adaptation. But we also know from the theory of the 'second best' that the consequences of a further regulation are not given.

One advantage of a tax over other regulations is the revenues it generates. In this connection, it is often pointed out that an environmental tax can yield twice (see Chapter 12). First of all it corrects a market failure, which provides increased welfare. Then the revenues allow reduction of taxes and charges which create efficiency losses, providing the second dividend. As we have already pointed out, the harm caused

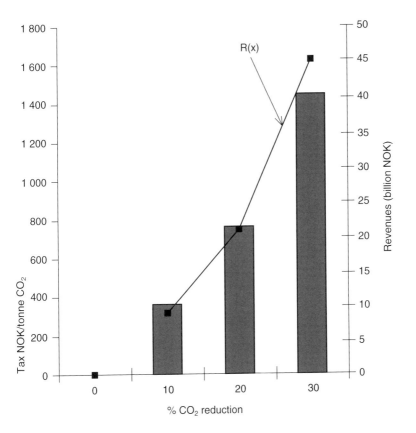

Note: The curve shows the CO_2 tax (NOK/tonne)

Figure 22.1 CO_2 tax and revenues at varying levels of reduction by emissions of greenhouse gases is probably something that lies far ahead of us, and is not included in our model. In this way, the first effect is not included in the model. The significance of the tax revenues should not be underestimated, however: (see Figure 22.1). Håkonsen and Mathiesen (1994) found in their model that, in the case of a 20 per cent reduction in CO_2 emissions in 2000 compared to the reference path, the revenues generated by the CO_2 tax needed to achieve such a reduction would be in the order of NOK 25 000 million. This is approximately equal to half of the revenues generated by employers' contributions in 1995.

Håkonsen and Mathiesen (1994) also found that the costs of a CO_2 reduction could actually also be negative – that is, that it could be socio–economically profitable to reduce emissions within certain limits when the economy was fundamentally inefficiently adapted in a number of areas, and that application of CO_2 taxation could affect adaptation in a direction which reduced efficiency losses. They proved, as has also been shown in a study for the USA, that the way the economy is modelled is very significant, especially how the revenues generated by the CO_2 tax are used and whether damage caused by the gases emitted in tandem with CO_2 is taken into account. Håkonsen and Mathiesen studied the following alternatives:

(a)   Transfer of all CO_2 revenues as a lump-sum grant to all households.
(b)   All CO_2 revenues are used to reduce employers' contributions.
(c)   Traffic injuries and health damage which results from the 'non-greenhouse' gases are taken into account when computing welfare.
(d)   Subjective 'injury' from these sources is also taken into account.

Figure 22.2 shows trends in a welfare index as a result of reduced CO_2 emissions in the four descriptions of the economy. The index is an aggregate of private sector product consumption and leisure, U(C,L) and is equal to the index from which the household's product and leisure demand is deduced. The index is far from being an ideal welfare indicator but it is, at least, an improvement compared to more traditional GNP indexes which ignore demand for leisure. In the (c) and (d) scenarios the savings in 'repair costs' for people and equipment are also taken into account. The difference in quantified welfare is appreciable. While that index which is based on the premise that the revenue effect is 'ignored', that is, index (a) indicates that a 20 per cent reduction will

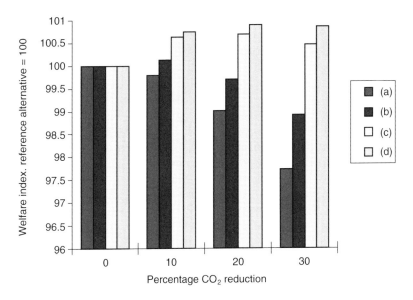

Figure 22.2   Welfare index for four scenarios and at varying levels of CO_2 reduction

cost 1 per cent in the form of lost welfare (in other words, an equivalent result to conventional studies of such reductions), the dividend from reduced employers' contributions (b) will reduce the cost to 0.3 per cent. If we also take into account the fact that reduced CO_2 emissions lead to reduction of other gases and of motor vehicle traffic, and that these sources also cause minor damage (c), the figure shows a total dividend of 0.7 per cent. In this way, CO_2 stabilisation appears as a 'no-regrets policy' – that is, a policy which should be implemented and adapted over a decade independent of the CO_2 goal. If we further take into account the subjective assessment of reduced traffic noise and improved health, the dividend is even greater (Brendemoen, Glomsrød and Aaserud, 1992).

The point is not to claim that a unilateral CO_2 stabilisation is really a 'no-regrets policy' for Norway, but the opposite impression (which is the result of any number of analyses – that is, that stabilisation leads to a socio–economic cost) cannot be taken as given, even if it does seem to be widely accepted. We would also claim that the quantified cost is a result of insufficient computation. The ring effects which are included in (c) and (d) are actually underestimated. We should also point out that the conditions which are included in the analysis are far from the only ones which are relevant in this connection. The costs and problems connected to change, which were discussed in Chapters 18–20, have been played down. Nor have the dividends of reduced emissions for future generations been taken into account.

Table 22.1 shows various consequences of two alternative utilisations of tax revenues. Reduction in employers' contributions yields lower wage costs, which leads to increased demand for work and therefore wage increases. Employers' contributions are reduced from 14.3 per cent to 6.5 per cent on average, and the real wage increases by 3.8 per cent compared to the reference path. The transfer is therefore a modest 0.5 – in other words half of the tax reduction goes to increase wages. The parliamentary employment committee assumed that the transfer would be 0.8 (NOU, 1992: 26). Mathiesen (1992) argues against this high estimate and Kolsrud and Nesset (1992) estimated a substantially lower figure of 0.2. (In other words, this is a question where there is considerable difference of professional opinion.) This would lead to reduced leisure and increased employment opportunities. A supply flexibility of 0.3 is assumed. The sensitivity analysis shows little variation for the parameter values between 0 and 0.6, with great variation for a model with endogenously determined employment opportunities from one with exogenously determined opportunities.

For our purposes, it is worth noting the difference in total employment of at least 2 per cent between (a) and (b). (b) stimulates increased activity and consumption. Industrial activity, seen as a whole, increases and the labour-intensive sectors gain from the change. These are also mainly the sectors which have the lowest emission levels. The change in fiscal policy from labour taxation to environmental taxation changes profitability patterns for industrial sectors. Emission-intensive sectors see reduction in profitability, while 'clean' sectors see improvement. The increase in industrial activity also increases the combustion

Table 22.1   The effects of a 20% reduction in CO_2 emissions in 2000

	Lump sum transfer to Households	Reduced employers' contribution
Taxation revenue	23mrd	25mrd
Product consumption	−1.9%	0.2%
Leisure	2.0%	−2.0%
Utility indicator	−1.00%	+0.30%
Employment	−1.1	1.1

of fossil fuels, also increasing emissions somewhat. For this reason, the model computes that an increase of around 5 per cent in the CO_2 tax is necessary to counteract this emission increase, which explains why taxation revenues are somewhat higher in alternative (b) than in alternative (a). These model results appear to give a larger and significantly nicer result than the MSG computations. While the five MSG scenarios over a 40-year period generate hardly any surplus revenue to reduce, for example, employers' contributions, these computations make a lot of room for reduced tax distortion at around the same percentage of CO_2 reduction as the MSG scenarios in 2000.

It is quite possible that these seemingly very different results from the models are not at all at variance which each other, if the results and assumptions are examined more minutely. It is clear that the MSG models examine only the GNP and impacts on consumption, while these computations examine alternative extended welfare goals. MSG and MODAG's reaction coefficients are also econometrically estimated from historical data, while the coefficient values in the SNF model are selected on the basis of what SNF researchers regard as 'reasonable' for these computations. The sector divisions for the most polluting activities are significantly different in the models and SNF has shown that the results of the computations are sensitive to choice of sector division. Finally, and possibly also most importantly, we must take into account the fact that these SNF computations apply for 2000, while the MSG computations examine 2030. Both demographic trends, and the heavily decreasing significance of petroleum revenues in the years approaching 2030, make the tax distortion potential of CO_2 taxation significantly less in 2030 than in 2000. It should also be pointed out that the CO_2 scenario in the MSG computations showed an appreciable revenue surplus in the first 10–20 years (see Chapter 18) which could have been used in this period to return a significant reduction, for example, employers' contributions. This indicates that if we had compared MSG results for 2000 with the SNF results, the deviations in this area would be significantly lower, and the conclusions about the potential for replacing employers' contributions with CO_2 tax revenue for the public sector mainly in agreement.

22.3   A comparison with estimates from EU states

The opportunities and limitations of 'green tax reform' have been examined in more detail in Denmark (Clemmesen, 1995). The main conclusion is that this could yield considerable environmental gain, as the same time as there is no need for any financial loss to be incurred; there may even be small but positive employment gains. The Danish study emphasises that the tax erosion effect may undermine the suitability of many of the environmental taxes as a stable source of public revenue in place of income tax or VAT. This is illustrated by a not unrealistic case where demand for the environmental good falls by 10 per cent when the price increases by 10 per cent as result of an increase in environmental taxation. If the environmental tax forms around half of the market price of the product, then a doubling of the environmental tax will increase the tax revenue by only 33 per cent because the tax foundation is reduced by 33 per cent when the price rises and demand falls. If the tax is doubled again from the new and higher level, the tax foundation is reduced further and the tax revenues will increase by only 20 per cent. In general, it is possible to say that the tax revenue effect is greater the less price flexibility there is in demand.

The Danish study examined the effects on employment of introducing a CO_2 tax of NOK 200 per tonne of CO_2 during a four-year period, compared with a revenue-neutral reduction in other taxes and charges. The following effects were identified:

(a)   A direct reduction in industrial competitivity with an associated loss of public revenue and a 1 per cent increase in unemployment (an increase of 29 000 unemployed after five years).

(b)   Reduced employers' contributions providing improved competitivity through lower wage costs. Employment increases by 24 000 and tax revenues increase.

(c)   To a certain extent, energy is replaced by labour and capital and a further 9000 jobs are created so that the net effect is a job increase of 4000 with a simultaneous improvement in the environment.

The costs of the reform and the health benefits have not been computed (Clemmesen, 1995).

While this Danish measure will generate around NOK 2100 million (or around 0.2 per cent of GNP, so that it becomes possible to reduce distortionary taxes on labour) it is not necessarily certain that there will be any significant effect on employment in the long run. As we also pointed out in Chapter 12, the total tax burden on consumers is not changed when income tax decreases as much as the environmental tax on consumption increases. Wage earners will then tend to demand compensation for the increase in environmental tax, and industrial wage costs will remain unchanged. Exactly the same reaction is expected when social security contributions are reduced. The conclusion must then be that environmental tax levied on households will not result in a net job creation, but the environment will benefit. Only in cases where environmental taxes are levied directly on industry, while at the same time refunding the extra cost through reductions in employers' contributions or as a lump sum, will companies gain access to cheaper labour and be fully compensated for the increase in environmental taxation.

Since only 30 per cent of the tax revenues come from industry, the study concluded that the positive impact on jobs would be very modest, while the positive impact on the environment was greater than at implementation of the measure. The study also pointed out that as environmental taxation was, on average, distribution-neutral, this sort of tax reform must be supplemented by increased marginal taxation on incomes and property in the interest of the total distribution effect.

There is considerable scepticism in Germany as to the political opportunity for introducing CO_2 taxes to an effective extent in the major polluting industries. For this reason, alternative measures with a greater probability of implementation are being developed (Luhmann, 1995). This alternative means that companies may not deduct for energy costs when calculating their pre-tax profit. In compensation, companies are allowed to deduct a fixed amount which is equivalent to energy costs in the case of energy-efficient adaptation, thereby stimulating industry to invest in lowering energy consumption. No environmental or financial consequence analyses of the proposal have been made.

DRI's study of the possible double dividends from EU environmental taxation (DRI, 1994) contains a comparison of a reference path and an alternative where all of the environmental externalities are internalised, using mainly economic measures (Pigou taxes) between 1990 and 2010. The study concluded that increasing environmental taxation by around 1.7 per cent of the total GNP in the six largest EU states might provide enough to reduce employers' contributions or income tax by an equivalent amount, making labour cheaper. Computations in which employers' contributions are reduced indicate that, in addition to appreciable environmental gains, there is a certain increase in the rate of growth of GNP and that, by 2010, employment will have increased by a total of 2.2 million jobs compared with the reference alternative. This is equivalent to a reduction in unemployment of 1 percentage point. There are also corresponding effect studies from other countries (for example, Brinner *et al.*, 1991) which show similar double dividends, and which therefore support the theoretical analyses and empirical results that we have found in the Norwegian case in Chapter 12.

Great care should, however, be shown in transferring the quantified conclusions from EU states, with their high levels of unemployment, to Norway whose unemployment level is, compared to the EU, relatively low. It is not unrealistic to believe that the employment effects are less when the unemployment is low to begin with. Another factor is that pollution and accumulated problems are higher in the EU than in Norway, indicating that this necessitates much higher local environmental taxation (apart from the CO_2 tax) than in Norway. Against this background, the increase in public revenues of 1.7 per cent of GNP which could result from the internalisation of EU's environmental taxation could be interpreted as an upper limit for Norwegian conditions if we accept the results of the DRI study. This means that if this logic holds, then the highest reduction of Norwegian employers' contributions without being detrimental to public sector (local and national) finances would be less than 1.7 per cent of GNP. 1.5 per cent of GNP, in present terms would be around NOK 10 000 million and this is around 20 per cent of state income from employers' contributions. A 20 per cent reduction in employers' contributions would, as the first stage in an adaptation, mean a reduction of between 14 per cent and 11 per cent in the upper tax level regions (and equivalent percentage reductions for lower tax level regions).[1] It would appear that revenues from environmental taxation, and through them the opportunity to reduce other taxes and charges, are limited in Norway. There are, however, other additional opportunities linked to the extra return from environmental and natural resources which may, for example, be used to reduce employers' contributions. This is discussed in more detail in the next section.

22.4   Economic rent and extra return[2]

'Economic rent' is an old expression within the field of socio–economics. One, oft-quoted, definition is from Ricardo, who states that economic rent is *payment or compensation for use of resources which are available from nature in a given quantity, irrespective of efforts exerted by the individual*. Around the turn of the century, Henry George, the American socio–economist, created a popular movement which aimed at replacing all taxation by a plain tax on basic values of all types of land (see George, 1880). George argued from the view that land was naturally suited to a given use. In this case a suitably located piece of land would have a higher value than another, more remote, piece of land, independent of the owners' efforts, just because it was closer to an urban centre. If taxation is levied on economic rent then it would be a tax on an extra return which might be both efficient and fair.

As the economic rent is created by nature and social trends, not by the efforts of an individual, taxation of economic rent would not create the same distortion as tax on labour and capital. In addition, it has an added advantage, in that those who have to pay the economic rent tax have the ability to pay it.

The extra return on natural resources and land may have various origins, and the economic rent we discuss here is an important part of it. For renewable resources and land (both rural and urban) it is the closeness to the market and the difference in the natural fertility of the soil which leads to the extra return. It is possible to rank similar projects (for example, power stations) according to how cheap their power production is.

The extra return may also arise because a resource is not renewable, so that future revenues are derived by extracting it today. As if this were not enough, the various measures and regulations lead to extra value or monopoly gains for certain players in a resource market. All of these possible sources of extra return are called 'economic rent' (even if this is socio–economically not quite precise).

Because the economic rent is calculated as an extra return from what the natural and resource base can offer, economic rent is usually regarded in the same light as resource management. When we attempt below to place a value on such extra values as a basis for less distortionary taxation, it is important to be aware that a number of comprehensive regulations, concessions and contracts already stake a claim to major parts of the theoretical extra return from the resource in question.

Economic theory has shown that taxation of economic rent is an efficient and non-distortionary form of taxation. If the public sector harvests the economic rent through taxation, allocation losses of the sort incurred with income tax do not occur. For this reason, economic rent has come into focus as an interesting alternative to distortionary taxes – for example, on labour – so that it may be possible to attain employment dividends at the same time as avoiding allocation losses in the economy. It is only necessary to decide that this is a dividend which belongs to the community, it is harvested by the community using taxation and not by an individual landowner who is coincidentally located near projects and resource values which the community has created or has access to. The economic rent is undoubtedly a suitable tax object.

The uncertainty that we shall focus on in this chapter applies to:

- the strong political pressure groups which must be overcome if fiscal reform round economic rent is to be achieved, and
- what actual revenues can be expected to be harvested through taxation of various economic rents and which can form a basis on which to reduce taxation of labour.

22.5 The economic rent on raw materials' industries

An examination of the economic rent for power stations will easily illustrate what we really mean by 'economic rent'. The costs of development and operation of hydropower stations vary as a result of natural conditions. This is harvested in those projects which, because of natural conditions, are able to be developed at a lower cost than the market price. The economic rent will be harvested through the concessionary use of watercourses. Before power sales became open to market forces it would not have been relevant to discuss taxation of economic rent because the economic rent was distributed to purchasers via low prices. The Norwegian Energy Act (Ot. prp. No. 57, 1989–90) determines that market prices must be attained in the power sector and that the profit must become visible.

Computations (see Hervik and Hauge, 1995) have been made which show that if power prices can find a new equilibrium equal to the capacity costs for new energy (NOK 0.41 KWh to household consumers) then today's economic rent would be as much as NOK 7000 million (while the alternative with production charges is around NOK 3000 million). Depending on whether the door is opened for energy exports or the energy sector is protected, the economic rent will increase (with export) or be reduced (if the sector is protected). In the Long-term Programme (LTP) 1994–7, the economic rent was calculated as being somewhat lower than NOK 6000 million per annum. When compared with today's taxation of the energy sector, a change to harvesting the economic rent would mean that we can increase tax revenues by around NOK 4000 million each year.

The most important institutional barriers against full introduction of taxation of economic rent in this sector are linked to the present system of providing low-cost power to power-intensive industries. Energy prices are used as a political measure in the maintenance of settlement and employment in outlying regions. The difference between the low economic rent that we see in the hydropower sector today (it is far less than half the theoretical economic rent) and the computed theoretical economic rent provides us with a goal which shows what is covered by regional policy costs in this method of disposal of economic rent.

The petroleum sector

The resource interest from the petroleum sector has varied widely as result of large price variations (see Figure 3.1., p. 39) and were, when we defined them as return over and above ordinary return on capital, calculated as being NOK 38 000 million in 1993 (Lurås, 1994).

There is a tradition in the petroleum sector for harvesting the resource rent with an economic rent tax of 50 per cent in addition to normal company tax (which is 28 per cent), leading to high tax revenue in a very profitable field. The problem with the system has been adjusting tax in line with changes in oil prices, so that the specific tax will not apply just in the fields which are the most profitable.

The institutional problem of a tax which intends to harvest all the resource rent is linked to the high level of uncertainty involved in the entire North Sea petroleum industry. It is difficult to determine the impact of this uncertainty, the impact of the great variation in oil prices and how to handle the threat from foreign oil companies to withdraw from the North Sea if oil taxes increase. Company taxation will vary widely, and there will be periods of low prices that may present an argument for lower taxation. Historically, it is probably a fact that a certain proportion of the resource rent in the North Sea has benefited the foreign oil companies. At present, we do not know the exact size of the rent, although we should have enough basic data to estimate it. Although we can manage the technology, it is important to have many players in the North Sea to ensure a satisfactory risk spread. During periods of low oil prices, it is necessary that some of the risk is spread by allowing some of the resource rent to benefit the oil companies, rather than the community through a tax on the resource rent or economic rent.

In 1994 the tax revenues levied on the oil companies amounted to NOK 21 000 million. Income from direct state financial involvement made up the remainder of the nearly NOK 40 000 million which was the total state revenue from the petroleum sector in 1994.

Securing incentives for further exploration must be weight balanced against harvesting as much as possible of the resource rent for the Norwegian community. One alternative could therefore be to sell the fields short and then collect the resource rent on the basis of these future sales. This may provide a reasonable distribution of the risk between the various players in the Norwegian sector, while at the same time making socio–economic efficient decisions. Even if the oil companies threaten to withdraw from the North Sea every time taxation of the economic rent is mentioned, it is continually proved at present that the incentive is sufficient because all the companies are still applying for concessions. The tax on economic rent could be organised as a computation of the remaining net income (*minus* a reasonable return on the capital, including the risk premium – 10–12 per cent) from the revenue. This figure should be harvested as tax on the economic rent, and will probably, in the long term, increase revenues. The other challenge for fiscal design is to provide incentives for extracting more oil from marginal fields, providing better exploitation of the resource base.

Fishery resources

The harvesting of Norwegian fishery resources is, at present, regulated by quotas. Fishing boats are granted concession to fish a specific quota. Until the end of the 1980s, the fishing industry received considerable state subsidies; these have been reduced heavily over the last few years, falling from NOK 2000 million as late as 1989 to NOK 200 million in 1994. The problem is overcapacity, so that something that should have allowed Norway to collect a tax on the economic rent has instead received subsidies. If the state had opened an exchange for quota sales in which all fishermen could bid, the economic rent would become visible as a desire to pay for the rights to harvest the resource. At present, the quotas are distributed among a large number of fishermen who, as a block, form an overcapacity which makes it difficult to fish profitably under the present system. Given a set of trends which mainly reduce capacity and subsidies, we can enter a new phase which gives Norway optimum capacity in a much smaller fishing fleet. Norway would then be able to harvest an economic rent to the community once more. The economic rent on fishery resources is estimated at nearly NOK 2000 million if the fleet capacity and costs were optimally adapted (Hanneson, 1991).

The most important institutional barrier within fishery management and structural reform in this industry is that most of the fishing industry and associated jobs are situated in outlying regions of Norway.

22.6 The economic rent and property tax

The general debate around a property tax has occupied a central position in tax reform. Some politicians have focused specially on the distribution problems which are inherent in designing tax assessment regulations. In principle, a property tax may assume the character of an economic tax to cover public expenses involved in investment in infrastructure, although this is an aspect which has largely gone unnoticed in the debate.

Two examples from rural and urban Norway, respectively, may illustrate this. Suppose we build a bridge costing NOK 800 million to an island of 1300 households (both the Hitra and Magerøy bridges are examples, one of them somewhat better, the other somewhat worse than our mean figures indicate). We assume that it will not be a toll bridge, but that the community allows the entire dividend to result in better accessibility for the inhabitants, so that their property values increase as a result. If we deduct savings in annual costs for ferry fares, and payment for industrial and tourist use, then the annual cost of the project itself, which must be covered by increased economic rent per household, will be over NOK 30 000. This means that if the project is to be profitable, the economic rent would have to increase on average by NOK 30 000 as an expression of increased accessibility. On average, the inhabitants would have to be willing to pay an increased property tax to the tune of NOK 30 000 to guarantee profitability.

If those who made the decision for the project go-ahead also had to approve the property tax increase to finance this public project, such projects would never take place because the inhabitants would never accept so large an increase in property tax as NOK 30 000. This also reveals that this type of investment in infrastructure may, in many cases, not be good regional policy. On the contrary, it is so expensive that none of the affected parties would see any benefit from it. It would be more easy to see that more moderate local projects would give less traffic growth; the result would be lower public costs and a better basis for sinking the levels of the present distortionary labour tax. Our point is that, given taxation on the economic rent and another distribution of functions – in which, for example, local property taxes are used to finance projects – decisions regarding local bridge or tunnel projects would be made on the basis of a more direct cost-responsible principle, which would result in a number of expensive infrastructure projects never coming to fruition.

In urban areas, public investment in the infrastructure could reduce land scarcity and create city development by improving accessibility to the heart of the area. This would then have the same effect on the economic rent in cities as when invested in rural projects. The urban property market is influenced by this and very many other conditions, making it difficult to estimate the link between development of infrastructures and increases in property values. Even without finding exact values for these links, high property values in central urban districts would be linked to historical infrastructure investment which make it possible to improve the performance of the city's functions. Increased land scarcity in central urban areas – not least resulting from transport systems occupying large areas – could therefore be ameliorated. If there are queues in city centres which have not been internalised in transport prices, the result could easily be overinvestment in the infrastructure of suburban areas (Solow and Vickrey, 1971). If efficient use of resources is to be achieved, the cost–benefit ratio must be high if land is preferred for roads rather than for industrial uses. Depending on the design of the queue-cost function, the cost–benefit ratio should be larger than four if it is to be socio–economically viable to develop the road network as a result of these imperfections in the land market.

The authorities can levy a tax on economic rents based on the marginal cost of investing in improved access in city centres. This would probably result in expensive tunnels or bridges, or expensive multistorey car parks, which would make the land scarcity visible. This is reinforced by environmental demands and traffic solutions which together lead to high capacity costs. Without a system of payment for marginal capacity costs through road pricing which reflects the costs of improved access, traffic will increase and create further capacity problems. If we now choose to price queues based on this principle, we may also have a method for cashing in the economic rent. We now pay an amount which is equivalent to the marginal crowding costs, which also reflect capacity costs and land scarcity in city centres, to use the infrastructure. This will not lead to an equivalent increase in the economic rent for new construction projects, because better roads are paid for by the road pricing system.

If this price level is established, the economic rent will be harvested from the other (previously realised) road investments which are now priced at a level equivalent to the marginal queue costs. Road pricing can now cover marginal queue costs as well as other environmental costs which follow from the heavy weight of traffic and the environmental strain which results from urban traffic, and will be a measure for limiting the its growth. It will then be regarded as a 'green tax'. There are clear parallels here to the energy sector. Publicly financed rail construction and subsidies to various forms of public transport will extend and develop city-centre functions. This will lead to increases in the economic rent which could have been used to finance infrastructure through a property tax for landowners who indirectly have gained benefits from the increased economic rent created by the publicly financed city-centre development. Institutionally, experience often shows that city-centre landowners are those who engage themselves most strongly against such forms of road pricing. This is, of course, because they will undermine the value of city-centre property. Our point is that precisely this tax will correct previous distortions which enabled landowners to benefit from a community-created economic rent which was exempt from taxation. There are already toll plazas around all major Norwegian cities, although those who live within the toll zone do not have to pay for use of scarce street space. The authorities should also increase prices during rush hours, as a more practical approach to taxation on economic rent.

Fuel prices cannot reflect queue costs in city centres. They will also deviate from costs of road construction in the more peripheral districts. In a price system where fuel taxes must take care of 'mean costs', road pricing must be a supplement which corrects for the effects of crowding. In the more peripheral districts, they must reflect the economic rent and road construction costs in urban areas.

Today, property tax is a municipal tax governed by state rules. In addition, the advantage of being an owner-occupier is part of the Norwegian tax system, making it also a state housing tax. For professionals, there is agreement that both forms of taxation are little used in Norway and that, historically speaking, the tax rules have provided an incentive to overinvest in housing and property. Norway has had a generally low level of taxation on housing capital. While in 1992 property tax was around 3 per cent of the total taxation revenue in Norway and Europe, the same figure was as much as 10–11 per cent in Australia and USA. Companies also benefit from the low property tax in most municipalities, meaning that low property taxes are a production factor.

Given a perspective of neutral design of the fiscal system, it would be correct to tax property and housing more and labour less. Our perspective has been to discuss taxation of the economic rent or property tax as an efficient form of finance of public production and increase taxation of the economic rent for those who benefit from such public (local) goods. The public sector has significant net costs connected to the infrastructure in general, and especially for production with high fixed costs and falling average costs.

It is not immediately easy to identify the direct link that this would have to 'green tax reform'. As property tax is directly linked to land, it will reduce the growth in properties if such a tax is introduced. This will initiate a better exploitation of housing space in the long term, and lead to less space-intensive production. In the second place it will lead to a decrease in heating costs and energy consumption, giving a positive environmental spin-off. We do not know whether transferring income use from housing to other consumer goods (such as travel and entertainment) will lead to increased strain on the environment; it will be possible to protect oneself against this only by generally internalising costs of environmental harm in the market prices through taxation.

Finally, we shall indicate the potential income of the tax reforms we have discussed here linked to taxation of economic rent on property, starting with the aspects connected to crowding and regional effects. In Norway's largest cities today crowding taxation on a level which would be necessary to correct for it would provide only a maximum of NOK 1–2000 million. If we tax the economic rent on regional road projects, we could make an annual saving of a maximum of NOK 0.5–1000 million by investing less in poorly profitable regional projects. Crowding taxes will limit the need to invest in cities and urban areas so that it would be reasonable to operate with a total maximum saving in the roads sector of nearly NOK 3000 million with these forms of taxation of economic rents. At the same time, fuel taxes were more than NOK 3000 million in 1994, and some of this 'tax piracy' would be eroded away when it became more expensive to use roads. In other words, no strong increase in tax revenues will be gained (or taxes saved) which could form the basis of an appreciable reduction in employers' contributions of around NOK 50 000 million.

Another natural resource, which gains ever-greater commercial significance, is nature, which means a lot in the choice of Norway as a tourist destination. The Institute of Transport Economics' holiday survey for 1994 calculated that 1.1 million foreign tourists stayed an average of 11 guest nights during the three summer months, a total of 12 million guest nights. 335 000 foreign vehicles contributed to these visits, including 3500 camper vans and 2500 caravans (*Aftenposten*, 15 May 1995). The net income created by tourism in Norway is also a sort of economic rent on environmental capital, as this is what attracts the tourists to Norway. There are many ways to capitalise on this economic rent through taxation – as in, for example, Switzerland, where tourists pay a fixed car tax when they enter the country. Norway has capitalised on the economic rent to a certain extent through our relatively high fuel taxes and by payment in the many toll plazas and ferry connections. Previously, ferry traffic employed summer rates to recoup many of the high costs which tourist traffic (which to a certain extent determined capacity on many crossings) inflicted. The system was scrapped in 1986 as a result of strong local opposition to what was regarded as a special tax on rural Norway. Later surveys have not indicated any possible link between tourist flow and the scrapping of the system (which is confirmed in surveys from other countries). Because the local population use discount cards (for example, coupon cards), avoiding the summer rates, it was mainly city dwellers, locals who travelled little and foreign tourists who paid this extra price.

Professionally speaking, there was no need to scrap the system, which was not at all a special tax on the population of rural Norway, although political expediency caused its demise. Consider the introduction of a system of capitalising on the economic rent of nature from car tourists, through an extra payment which the travellers paid in toll plazas (and at ferry crossings) equivalent to an increase of 50 per cent of today's rates. This would be equivalent to the pre-1986 system which would now contribute almost NOK 400 million in extra public revenue without any quantifiable loss of tourists. There should not be any obstruction to further discrimination towards such inflexible demand.

Over the last few years, a number of countries have introduced heavy price discrimination towards foreigners who come to enjoy their scenery. Bhutan has heavily increased the entry price per visit – without the country becoming a less attractive tourist destination. Nepal has also heavily increased its expedition charges, and there is still a waiting list of climbers. The Galapagos islands increased its tourist charges 20 times without any reduction in tourist traffic. Safari charges in African animal reservations are another example, and in China all the prices of visits to tourist attractions were increased significantly without affecting demand to any great extent (Hansen, 1993b).

The point is that this taxation of economic rent, which would be felt by the local population if it was levied on them, represents a one-off payment for travellers, and the amount will, in any case, be only a very modest part of the tourist's total holiday budget. Tourist traffic is also in many areas a dimensioning factor for road projects and the infrastructure, so it would seem only reasonable that the tourists themselves pay for the projects, and not the other Norwegian taxpayers elsewhere in Norway (for example, in the Magerøysund and Hitra connection examples).

If we examine property or housing taxation more closely, we see that we could achieve a significantly extended tax base. Property tax at the local authority level would lead to a correct distribution of functions to finance local common goods where the production decision can be made locally. Those who can enjoy the benefit must also pay for it, and a reasonable local consideration of costs and benefits is achieved. A local property tax would form today a minor tax base compared with financing local common goods which have indirect significance for property values, and therefore for economic rent. Water, sewage and refuse collection charges are mandatory charges linked to residences and fully finance such goods as are found in all municipalities, in contrast to property tax. Subsidies to the public transport network can be interpreted as a typical example of a local common good which should be funded by property tax in addition to the general subsidy framework. This would give a basis for reduced labour taxation. Local bus routes, ferries and railways alone receive more than NOK 3000 million in subsidies at present, while the total property tax revenues for 1993 were NOK 2700 million and a property tax was levied in 200 municipalities.

The best possible exploitation of the resources of the community requires that taxation of the various savings schemes for housing capital, must be designed in such a way that the saving which provides the highest pre-tax return also gives the highest return after tax if the tax is not to influence the choice of savings object. The same principle should also apply between investment in housing capital and other investment. A 28 per cent tax is generally levied on return on capital, in addition, a wealth tax can be levied on money placed in bank accounts or in shares. The size of tax on housing capital is based on the taxable value of the house, a baseline reduction and a general addition to income of 2.5 per cent of the value, rising to 4.5 per cent for the most expensive house (a progressive housing tax).

If we compare the margins in an investment of NOK 100 000 in housing with other investment, the other investment yields approximately 6 per cent interest, which is taxed at 28 per cent, so that 1.7 per cent of the NOK 100 000 would be paid in tax. If we assume that 6 per cent is the alternative value of the same investment in housing, then 2.5 per cent would be added to other income and a marginal tax of around 50 per cent would lead to a collection of a 1.2 per cent tax, meaning that the fiscal system favoured investment in housing. If we used 4.5 per cent as the rate for the income asset for an owner-occupier, we turn the conclusion on its head. The Aarbakke Committee (NOK 1989: 14) recommended that housing values should be linked to market value, although we are far from this goal today because we have tax valuations on housing where no more than 5 per cent of the housing market is affected by the 4.5 per cent rate.

Under the present system, we have a fiscal system which provides incentives for investment in expensive housing. The primary environmental consequences of this are that more and more land is used for housing, with resulting higher energy and material consumption from the building of 'too large houses'. (Whether the alternative consumption would be used for energy and materials is difficult to know.) If we change to basing housing valuation on the actual market value, a greater proportion of the total housing mass would cross the 4.5 per cent boundary, yielding increased tax revenues of NOK 4–5000 million, and reduced demand for large houses.

The tax on the asset of being an owner-occupier yields revenues of around NOK 1500 million. VAT of 23 per cent is levied on consumer goods. Services are exempt from VAT, although introduction of VAT on services, possibly at a lower rate, is being discussed. Housing capital can be interpreted as providing annual housing services. This is a part of private consumption which would be exempt from general VAT, and is in general taxed at a low rate through income self-assessment. Interest costs can be interpreted as an annual payment for the use of one's own residence. This expense is not liable to VAT – on the contrary, there is a benefit from the tax deduction of 28 per cent. An upgrade of the taxable value of a residence would mean that every household would pay an average of NOK 4–5000 in housing tax, while enjoying an equivalent reduction in income tax. This would also put housing tax on an equal footing with other goods.

As we can see, the total of all of the theoretically possible economic rents which we have discussed here, and which have not been collected by the authorities, to no extent approaches the NOK 50 000 million collected in employers' contributions. Even if, in time, agricultural subsidies were reduced, as has been indicated in the 'GATT alternative' or 'in further liberalisation' (see Chapter 20) by NOK 3000 and 6000 million, respectively, there would still be a long way to go. The maximum feasible implementation would be a gradual system of fiscal change, where employers' contributions were reduced by between NOK

15 000 and 20 000 million – in other words, from the present 14.1 per cent (in the highest tax class) to somewhere between 8 and 10 per cent – in return for taxing various economic rents as we have described, and reducing various subsidies to resource-based industries, collecting an equivalent amount.

There is, however, some opposition to taxation of economic rent through property. This is mainly of a practical nature, although we also recognise the theoretical problems which arise:

1. If we tax increases in property value on the basis of a public project, then compensation must also be given for loss of value – for example, from increased noise from high-speed rail traffic.
2. It is difficult to identify precisely who attains increased economic rent, and the tax object is not precisely delineated and may be regarded as unfair.
3. Property is bought and sold, and when the tax is introduced, a previous owner may have already capitalised on the economic rent in the market.
4. The distribution problem with property and housing tax is often promoted.
5. It is difficult to use the market to distinguish the economic rent from any value that the owner creates through his/her own efforts. This may create incentive problems for public operation.
6. In certain cases, the economic rent may reflect future conditions (for example, where a new road is being built). In this case, it will be capitalised on only after many years, so that an owner who is not able to pay the tax today would experience a cash-flow problem. In such cases, the tax must be linked to an increasing base value, and can be capitalised only with the increase in base value after the tax was introduced.

Even with the change from tax on labour to resource use which we have discussed here, there would still remain a significant level of distortionary taxation on labour. There should be no problem with reducing a distortionary tax on labour until the point is reached where the marginal tax is equally distortionary for labour, capital and other production factors.

Part VI
Conclusions

23

The Achievement of Sustainability Goals in the Five Alternative Projections *

In this chapter we shall examine to what extent the goals and demands expressed through the indicators which it has been possible to incorporate in the macro-economic models (see Chapter 10) were met during the process. We shall concentrate on comparing the demands and goals outlined in Chapter 10 with the results outlined in Chapter 18.

23.1 The indicators

Table 23.1 sums up the indicators for sustainable economy that the model scenarios intend to study. The table shows the increase or decrease as a percentage compared with 1989, and also the goals. The goals are strict compared to trends in the Long-term Programme (LTP). The alternative scenarios can redeem only a very few of the indicators' goals. Scenario 5 – the alternative where a significant part of the increase in welfare is redeemed in the form of increased leisure – comes nearest to meeting the goals. Scenario 2 also emerges relatively well. It is, however, important to examine the size of the costs in the various scenarios compared to sustainability (how much environment you get for your money in each scenario). We shall return to this point later.

23.2 Choosing between the alternatives

The choice of best alternative will be determined by how the various indicators are weighted and how harmful the emissions are believed to be, while the harm is assessed against the economic consequences

---

* This chapter is based on Alfsen, Larsen and Vennemo (1995).

Table 23.1   Indicators, 2030 (Percentage change from 1989)

	Goal	LTP	Alternative 1	2	3	4	5
Energy consumption per head	−50	16	12	12	−2	1	−7
CO_2 emissions	−60	17	2	−19	−23	−13	−30
CH_4 emissions	−20	−4	−6	−5	−7	−7	−7
NO_x emissions	−50	−11	−19	−24	−28	−26	−34
SO_2 emissions	−90	−63	−62	−78	−75	−67	−76
NMVOC emissions	−75	−29	−51	−41	−58	−58	−62
Production of raw aluminium	0	36	42	−40	−18	42	−24
Production of art. nutrients	0	45	57	−69	−40	57	−47
Production of cement	−70	56	63	23	40	36	10
Production of timber	0	183	193	148	168	163	132
Nuclear, gas and wind power	0	+0	0	0	0	0	0
Consumption* of metals	−80	96	102	68	79	95	50
Consumption* of art. nutrients	−80	30	33	4	15	20	0
Consumption* of cement	−80	91	57	96	57	73	59
Consumption* of timber	−10	220	242	143	188	176	108

Note: *The definition of 'consumption' used here is all domestic consumption in Norway –
that is, production *plus* imports *minus* exports (measured per head of population).

which accompany the measures. The fact that most of the emissions, apart from CO_2 emissions, can be cleaned (that is, reduced in other ways than those shown here), must be taken into account.

Figure 23.1 shows the percentage change into emissions broken down in percentage change in GNP and consumption. In alternative 1, the CO_2 emissions are reduced by 3.1 per cent for each percentage point of reduction in GNP. The higher the figures in Table 23.1, the less it costs to reduce the emissions. It can be read from Figure 23.1 that reducing working hours is, in isolation, an expensive method of reducing emissions, although the benefit to workers of increased leisure is not taken into account. The question of whether the type of strategy chosen in alternative 5 is good or not depends, therefore, on whether people appreciate increased leisure.[1]

Alternative 5 is to be preferred if the CO_2 reduction is to be regarded independently of costs, as emissions are reduced most in this alternative. From an economic point of view, however, alternative 2 is to be preferred, because the CO_2 emissions are much reduced compared with what the community must pay in the form of reduced GNP and consumption. This alternative is also to be preferred if you take into account sulphate pollution, as a result of the relatively heavy reduction

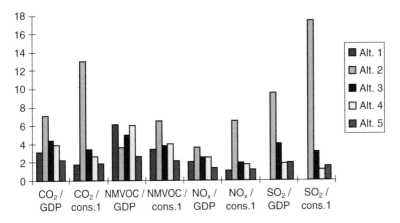

Figure 23.1 Emission elasticities with respect to GNP and consumption, percentage change from the reference path, 2030

which takes place in the activities of the power-intensive industry. Alternative 2 is also preferred with regard to NMVOC emissions, although only weakly in this case. Alternatives 1 and 4 are preferred when seen against the background of the NMVOC emissions – that is, the measures which are most clearly directed towards the petroleum sector, although cost efficiency does not vary highly in comparison with the other alternatives.

The indicators connected to aluminium, artificial nutrients, cement and timber change the most compared to the change in GNP or consumption in alternative 2. The least change is in alternative 4 which exempts the power-intensive industry from the increase in CO_2 taxation. If a great change in production and consumption of these products is the desired goal, then alternative 2 with its high, non-discriminatory CO_2 tax would be preferred.

As far as national wealth is concerned, none of the alternative scenarios distinguished themselves in a positive or negative direction. The impact on public budgets does, however, vary strongly. Alternative 5 distinguished itself adversely by its huge deficit in public budgets: a reduction in working hours must therefore be compensated for by increased taxation or reduced costs. The opportunity for a strong increase in tax revenues depends on whether there are good tax options. In Chapter 22, we pointed out some examples where it would be possible to tax non-mobile factors which might compensate for the unfortunate effects of alternative 5.

23.3 Conclusions

We have taken as given in the model analyses that sustainability can be measured by the set of indicators which we identified and discussed in Chapters 9 and 10. It is, however, clear that the model computations illuminate far from all the aspects of sustainable development. It seems in particular that many aspects of sustainability are closely linked to trends in primary industries. For this reason, we analysed these aspects particularly in Chapters 13 and 20.

It is difficult (not to say impossible) to give precise instructions about what is necessary to ensure sustainable development. We have discussed a wide range of theories and surveys which show the connection between the environment and various economic trends. The Project for a Sustainable Economy can in our estimation thus contribute to ensuring that national economic realities are included in the assessment and discussion of what we should understand as 'sustainable development'. At the same time, we have outlined some policy changes which may lead development in a more sustainable direction. Our hope is that the project can contribute to a more constructive level of debate and improved communication between those parties in society who, respectively, see it as their primary task to protect nature and the environment and who manage the Norwegian economy. We hope that this may form part of the basis for a debate about the creation of Norwegian policies in the decades to come.

Notes

Preface

1. Friends of the Earth–Norway is known in Norwegian as *Norges naturvernforbund* (NNV). It is Norway's largest conservation and environment organisation, with between 30 000–40 000 members. The Project for an Alternative Future (PAF) was originally an independent 'alternative' research programme established by leading members of Norway's NGO community with direct support from the Norwegian Parliament in the early 1980s. Its mandate was later transferred to the Research Council of Norway and ultimately transformed (in 1995) to a new Programme for Research and Documentation for a Sustainable Society (ProSus). The 'initiators' of the Project for a Sustainable Economy were Dag Hareide from NNV, Stein Hansen from Nordic Consulting Group AS, and William M. Lafferty from PAF/ProSus.

3 Towards an operationalisation of sustainability

1. Sustainability in the two illustrative scenarios below is so-called 'weak' (as opposed to 'strong') sustainability. These concepts will be discussed and clarified in Chapter 5.

7 Environmental valuation as a basis for environmental and resource management

1. Impact survey regulations are part of Chapter VIIa of the Norwegian Building Act (14 June 1985).

13 The environmental impact of deregulated agriculture

1. This section is based on Lothe (1995).

18 Comparision of government projections with environmentally adjusted projections

1. These energy sources formed around 30 per cent of the total energy consumption in 1989.
2. The revenues from state petroleum activities are exogenous, while the direct taxation and charges from oil extraction are endogenous.
3. The CO_2 tax measured in NOK per tonne can be converted to NOK per litre by multiplying the density of the oil (0.85 kg per litre) by the CO_2 content (0.00315 tonne CO_2 per kilo).

4. The current account balance is made up of the product and services balance (export *less* import of goods and services, and the unrequited transfers balance (net interest and grants to/from abroad).
5. In the alternative scenarios which presuppose reduced oil extraction, the remaining petroleum reserves are greater at the end of the forecast period. These extra reserves are presupposed to have the same value as the remaining reserves in the LTP. This must, however, be regarded as a low estimate, as the marginal costs of petroleum extraction are rising. In alternatives 2–5, a national CO_2 tax which does not affect the international oil price is presupposed. If the tax was imposed internationally, the petroleum wealth could be substantially reduced (see LTP 1994–7).

19 The presumed impact of CO_2 taxation on various sectors

1. This section is based on Mathiesen (1995).
2. This applies to the estimates from workforces, man-made capital, export prices and volumes, public sector consumption and budget shares in private consumption. (See also Rønning, 1994).

22 More efficient taxation from environmental policy

1. Norway has a geographically differentiated employer contribution tax.
2. This section is based on Hervik and Hauge (1995).

23 The achievement of sustainability goals in the five alternative projections

1. If the increase in leisure is valued as the mean post tax wage for wage earners, the consumption of increased leisure would be NOK 65 000 million in 1989 terms, in 2030. In comparison, the total fall in consumption (products and services) will be NOK 162 000 million.

Bibliography

3M (1994) *The 3M Environmental Progress Report*, Brochure (78-6900-6915-4), updated version available at website 'http://www.mmm.com/profile/envt/>.

Aasness, J., T. Bye and H. T. Mysen (1995) *Welfare effects of emission: taxes in Norway*, Statistics Norway, Research Department, Discussion papers, Statistics Norway, no. 148.

Alfsen, K. (1993) *Green GDP – Do We Need It?* ProSus *Report*, 2 (December).

Alfsen, K., B. Larsen, and H. Vennemo, (1995) *Bærekratha Ønonomi – Noen alternative modellscenarier for Norge mot år 2030* (Sustainable Economy – Some Alternative Model Scenarios as We Approach 2030), PSE Report, 15 (August).

Andersson, L. and T. Appelquist, (1990) Istidens stora växtätare utformade de nemorale och borene-morala ekosystemer. En hypotes med konsekvenser för naturvården, *Svensk Bot*, Tidskr. 84: 355–68.

Andrén, H. (1994) 'Effects of habitat fragmentation with different proportions of sustainable habitat: a review', *Oikos*, 71: 355–66.

Asheim, G. B. (1994a) 'Capital Gains and Net National product in Open Economies', *Journal of Public Economics*.

Asheim, G. B. (1994b) 'The Weizman Foundation of NNP with constant Interest Rates', Sosialøkonomisk Institutt, University of Oslo, mimeo.

Asheim, G. B. (1994c) 'Net National Product as an Indicator of Sustainability', *Scandinavian Journal of Economics*, 98: 257–65.

Asheim, G. B. (1995a) *Kompenserende Investeringer i en ressursrik åpen økonomi* TSE report 10, June 1995.

Asheim, G. B. (1995b) 'Økonomisk Analyse av bærekraft, Bærekraftig utvikling – om bærekraftens betingelser og utviklingens mål, Oslo: Ad Notam Gyldendal 1995.

Asheim, G. B. and K. Brekke, (1994) 'Sustainability When Resource Management has Stochastic Consequences', Arbeidsnotat University of Oslo.

Aunan, K., H. M. Seip, and H. A. Aaheim (1993) 'A model framework for ranking of measures to reduce air pollution with a focus on damage assessment', *Working Paper* 1993: 12, Oslo: CICERO (December).

Ballard, C. L. and D. Fullerton (1992) 'Distortionary Taxes and the Provision Of Public Goods', *Journal of Economic Perspectives*, 3.

Ballard, C. L., J. B. Shoven and J. Whalley (1985) 'General Equilibrium Computations of the Marginal Welfare Costs of Taxes in the US', *American Economic Review*, 75: 128–38.

Barro, R. J. (1974) 'Are Government bonds net wealth?' *Journal of Political Economics*, (November–December), 1095–117.

Bartlett, S. (1993) 'The Evolution of Norwegian Energy Use from 1950 to 1991', *Rapport 93/21*, Oslo: Statistics Norway.

Baumol, W. J. and W. E. Oates (1988) *The Theory of Environmental Policy*, Cambridge: Cambridge University Press.

Benestad, O., A. Kristiansen, E. Selvig, H. Westskog, L. Emborg, N. I. Meyer and L. Brinck (1991) 'Energi 2030 – Lavenergisecenarier for Danmark, Norge og Sverige', Oslo: Prosjekt Alternativ Framtid.

Berg, Å. and T. Pärt (1994) 'Abundance of Breeding Farmland Birds on Arable and Set-aside Fields at Forest Edges', *Ecography*, 17: 147–52.

Bernes, C. (1993) 'Nordens miljø – tilstand, utvikling og trusler', Nordisk Ministerråd, *Nord*, 11.

Bjerve, P. J. (1989) *Økonomisk planlegging og politikk*, Oslo: Det norske samlaget.

Børve, K., R. Gaasland, J. Brunstad, Ø. Hoveid, A. Huus, K. Mittenzwei and S. S. Prestegard (1994) 'Konsekvensvurdering av EU-medlemskap for norsk landbruk', *Rapport*, C-032-94, Oslo: NILF.

Bovenberg, A. L. and R. A. de Mooij (1992) 'Environmental Taxation and Labor Market Distortions', *Working Paper*, Ministry of Economic Affairs, The Hague.

Bovenberg, A. L. and R. A. de Mooij (1994) 'Environmental Levies and Distortionary Taxation', *American Economic Review*, 84: 1085–9.

Bovenberg, A. L. and L. H. Goulder (1994) 'Optimal Environmental Taxation in the Presence of Other Taxes: An Applied General Equilibrium Analysis', *Working Paper*, Department of Economics, Stanford University.

Bovenberg, A. L. and F. van der Ploeg (1992) 'Environmental Policy, Public Finance and the Labor Market in a Second-best World', *Journal of Public Economics*, 55.

Bovenberg, A. L. and F. van der Ploeg (1993a) 'Does a Tougher Environmental Policy Raise Unemployment?', *Working Paper*, University of Tilburg.

Bovenberg, A. L. and F. van der Ploeg (1993b) 'Green Policies in a Small Open Economy', *Working Paper*, University of Tilburg.

Bovenberg, A. L. and F. van der Ploeg (1993c) 'Consequences of Environmental Tax Reform for Involuntary Unemployment and Welfare', *Working Paper*, University of Tilburg and Amsterdam University.

Brakel, M. and M. Buitenkamp (1992) *Sustainable Netherlands – A Perspective for Changing Northern Lifestyles*, Amsterdam: Friends of the Earth – Netherlands.

Bramsnæs, A. (1991) 'Naturgenopretning – gamle og nye landskaber?', *Flora og Fauna*, 96: 85–92.

Brekke, K. A. *et al.* (1989) 'Petroleumsformuen – Prinsipper og beregninger', *Økonomiske analyser*, 5, Oslo: Statistics Norway.

Brendemoen, A. and H. Vennemo (1993) 'The Marginal Cost of Funds in the Presence of External Effects', Oslo: Statistics Norway, mimeo.

Brendemoen, A., S. Glomsrød and M. Aaserud (1992) 'Miljøkostnader i makroperspektiv', *Rapport*, 92: 17, Oslo: Statistics Norway.

Brendemoen, A., M. I. Hansen and B. M. Larsen (1994) 'Framskriving av utslipp til luft i Norge. En modelldokumentasjon', *Rapport*, 94: 18, Oslo: Statistics Norway.

Brinner, R. E., M. Shelby, J. M. Yanchar and A. Cristofaro (1991)'Optimizing Tax Strategy to Reduce Greenhouse Gases without Curtailing Growth', *The Energy Journal*, 12: 1–14.

Brown, L. R. (1990) *Fremskrittets Illusjon – State of the World 1990*, Oslo: Aschehoug.

Buchanan, J. M. (1960) *Fiscal Theory and Political Economy*, Chapel Hill, NC: University of North Carolina Press.

Buchanan, J. M. (1968) *The Demand and Supply of Public Goods*, Chicago: Rand McNally.

Buchanan, J. M. (1969) 'External Diseconomies, Correction Taxes and Market Structure', *American Economic Review*, 59: 147–77.

Buchanan, J. M. and G. Tullock (1962) *The Calculus of Consent*, Ann Arbor, MI: University of Michigan Press.

Budsjettnemnda for jordbruket (Budget Committee for Agriculture) (1993) 'Jordbrukets totalregnskap 1991 og 1992 og jordbrukets totalbudsjett 1993', Oslo.

Bye, B., T. Bye and L. Lorentsen (1989) 'SIMEN-Studies of Industry, Environment and Energy towards 2000', *Working Paper*, no. 44, Central Bureau of Statistics, Oslo.

Bye, T., Å. Cappelen, T. Eika, E. Gjelsvik and Ø. Olsen (1994) 'Noen konsekvenser av petroleumsvirksomheten for norsk Økonomi', *Rapport*, 94: 1, Oslo: Statistics Norway.

Cale, P. G. and R. J. Hobbs (1993) 'Landscape Heterogeneity Indices: Problems of Scale and Applicability, with Particular Reference to Animal Habitat Description', *Pacific Conservation Biology*, 1: 183–93.

Capros, P. (1994) 'Tax Reform within the EEC Internal Market: Empirical Analysis with Two Macroeconomic Modelling Approaches', in A. Heimler and D. Meulders (eds) *Empirical Approaches to Fiscal Policy Modelling*, London: Chapman & Hall, pp. 263–86.

Capros, P., P. Georgakopoulos, S. Zografakis, S. Proost, D. van Regemorter, K. Conrad, T. Schmidt, Y. Smeers and E. Michiels (1994) 'Double dividend analysis: first results of a general equilibrium mode (GEM-E3) linking the EU countries', in C. Carraro and D. Siniscalco (eds) *Environmental Fiscal Reform and Unemployment*, Dordrecht: Kluwer Academic Publishers, pp. 193–227.

Carlsen, A. J, J. Strand and F. Wenstøp, F. (1993) 'Implicit Environmental Costs in Hydroelectric Development: An Analysis of the Norwegian Master Plan for Water Resources', *Journal of Environmental Economics and Management*, 25: 201–11.

Carraro, C., M. Geleotti and M. Gallo (1994) 'Environmental Taxation and Unemployment: Some Evidence on the Double Dividend Hypothesis in Europe', *Working Paper*, GRETA Econometrics.

Christoffersen, K. (1995) 'Tidsbehov ved feltoperasjoner ved ulik feltutforming, areal og transportavstand', *ITF Rapport*, 60, Norsk Landbrukshøgskole, Ås.

Clemmesen, K. (1995) 'Employment Effects of an Ecological Tax Reform: Environmental Taxes in Denmark based on the Dithmer Report', paper presented at the Danish Board of Technology Conference on Ecological Tax Reform, Copenhagen (22 June).

Clench-Aas J., S. Larsse, A. Bartonova and M. Johnsrud (1989) 'Virkninger av luftforurensninger fra veitrafikk på menneskers helse', results from a study of the Vålerenga/Gamlebyen-area, Oslo, mimeo.

Commission of the European Communities (1993) 'Is There Anything New in the Concept of Sustainable Development?', *Working Paper*, CEC, Brussels.

Daly, H. (1992) *Steady-State Economics*, London: Earthscan Publications.

Daly, H. and J. Cobb (1989) *For the Common Good*, London: Green Print.

Dånmark, G. (1992) 'EøS og landbruket', *Landbruksøkonomisk Forum*, 1: 47–54.

Direktoratet for Naturforvaltning (DN) (1992a) 'Biologisk mangfold i Norge. En landstudie', *Rapport*, 5A, Trondheim.

Direktorate for Naturforvaltning (DN) (1992b) 'Biologisk mangfold i Norge. En landstudie', *DN-Rapport*, 92: 5, Trondheim.

Direktoratet for Naturforvaltning (DN) (1992c) 'Truede arter i Norge', *DN-Rapport*, 92: 6, Trondheim.

Direktoratet for Naturforvaltning (DN) (1994) 'Verdifulle kulturlandskap i Norge. Mer enn bare landskap', Part 4: Final report from the Central Committee, Trondheim.

Dixit, A., P. Hammond and M. Hoel (1980) 'On Hartwick's Rule for the Regular Maximin Path of Capital Accumulation and Resource Depletion', *Review of Economic Studies*, 47: 551–6.

Dover, J. A. (1991) 'The Conservation of Insects on Arable Farmland', in N. M. Collins and J. A. Thomas (eds), *The Conservation of Insects and their Habitats*, London: Academic Press.

DRI (1994) *Potential Benefits of Integration of Environmental and Economic Policies*, London: Graham & Trotman.

Ekstam, U., M. Aronsson and N. Forshed (1988) Ängar: *Om naturlig slåttemaker i odlingslandskapet*, Stockholm: LTS forlag.

El Sevafy, S. (1992) 'Sustainability, Income Measurement, and Growth', in R. Goodland, H. E. Daly and S. El Sevafy (eds) *Population, Technology and Lifestyle: the Transition to Sustainability*, Washington DC: Island Press.

ENCO (1993) 'External Costs of Fuel Cycles – National Implementations for Gas', Sandvika; Enco Memorandum.

Esbjerg, P. (1987) 'Insektslivets betingelser på danske landbrugsarealer', *Ent. Meddr.*, 55: 77–84.

Farmer, M. (1995) *Bærekraftighet i et økonomisk perspektiv*, PBØ Report, 8 (June).

Finansdepartementet (Ministry of Finance) (1993) 'Langtidsprogrammet 1994–1997', *St. meld. nr. 4* (Parliamentary White Paper), Oslo: Ministry of Finance.

Finansdepartementet (Ministry of Finance) (1992) 'Mot en mer kostnadseffektiv miljøpolitikki 1990–årene', *NOU 1992: 3*, Oslo: Ministry of Finance.

FNs klimapanel (1990) 'Globale klimaendringer', Rapport fra FN s klimapanel (IPCC) (Report from United Nations' Intergovernmental Panel on Climate Change), Oslo: Ministry of the Environment.

Fogelfors H., G. Bjørkhem (1986) Miljøeffekter vid använndning av energigrødor från jordbruket Solna: Statens naturvårdsverk.

Forsell, L. (1992) *EFs landbrukspolitikk*, Oslo: Universitetsforlaget.

Franklin, J. F. (1993) 'Preserving Biodiversity: Species, Ecosystems or Landscapes?', *Ecol. Appl.*, 3: 202–5.

Freeman, A. M. (1982) *Air and Water Pollution Control: A Cost–Benefit Assessment*, New York: Wiley.

Frivold, L. H. (1993) 'Kulturlandskap og forskjellige oppgaver', *LØF*, 2: 5766.

Fry, G. L. A. (1991) 'Conservation in Agricultural Ecosystems', in I. F. Spellerberg, F. B. Goldsmith and M. G. Norris (eds) *The Scientific Management of Temperate Communities for Conservation*, Oxford: Blackwell Scientific.

Fry, G. and A. R. Main (1993) 'Restoring Seemingly Natural Communities on Agricultural Land', in D. A. Saunder, R. J. Hobbs and P. R. Ehrlich (eds) *Nature Conservation, 3: Reconstruction of Fragmented Ecosystems, Surrey*. Beatty & Sons.

Gabrielsen, M. and A. Vatn (1988) 'Kanaliseringspolitikk og regional spesialisering', Report, 1, Landbrukspolitikk og miljøforvaltning, SEFO, Ås.

Geller, H., J. DeCicco and S. Laitner (1992) 'Energy Efficiency and Job Creation: The Employment and Income Benefits From Investing in Energy Conserving Technologies', *Working paper* for Environmental Defense Fund, Washington DC.

George, H. (1880) *Progress and Poverty*, New York: Appleton.

Gibson, C. W. D., V. K. Brown, L. Losito and G. C. McGavin (1994) 'The Response of Invertebrate Assemblies to Grazing', *Ecography*, 15: 166–76.

Gillebo, T. (1990) *Jordbruksareal i Norge – Utvikling og marginalisering*. Oslo: MiljØverndepartementet.

Glimskär, A. and R. Svensson (1990) 'Vegetationens förändring vid gjödsling och ändrad hävd. Trettiofemåriga skötselsförsök i naturbetesmark', Report 38, Institutionen för ekologi och miljövård, Sveriges Lantbruksuniversitet, Uppsala.

Glomsrød, S. and A. Rossland (1988) 'Luftforurensning og materialskader', *Rapport*, 88: 31, Oslo: Statistics Norway.

Goldemberg, T., B. Johansson, A. K. N. Reddy and R. H. Williams (1988) *Energy for a Sustainable World*, New Dehli: Wiley Eastern.

Golombek, R. and A. Raknerud (1995) 'Environmental Regulations and Manufacturing Employment: A Microeconometric Study on Norwegian Data', (28 February), Oslo: SNF, mimeo.

Goulder, L. H. (1994a) 'Environmental Taxation and the "Double Dividend": A Reader's Guide', *Working Paper*, Stanford University.

Goulder, L. H. (1994b) 'Effects of Carbon Taxes in an Economy with Prior Tax Distortions: An Intertemporal General Equilibrium Analysis', *Journal of Environmental Economics and Management*, 29(3): 271–97.

Goulder, L. H. (1994c) 'Energy Taxes: Traditional Efficiency Effects and Environmental Implications', in J. M. Poterba (ed.), *Tax Policy and the Economy*, Cambridge, MA: MIT Press.

Grime, J. P. (1979) *Plant Strategies and Vegetation Processes*, New York: Wiley.

Haavelmo, T. (1993) *Økonomi, Individ og Samfunn*, Oslo: Universitetsforlaget.

Haavelmo, T. and S. Hansen (1992) 'On the Strategy of trying to reduce economic inequality by expanding the Scale of Human Activity', in R. Goodland, H. Daly, S. el Serafi and B. von Droste (eds), *Environmentally Sustainable Economic Development: Building on Brundtland*, Oslo: J. W. Cappelsen/UNESCO: 41–51.

Hågvar, E. B. (1991) 'Biologisk kontroll av skadeinsekter', Faginfo, SFFL, *Report*, 23: 81–90.

Håkonsen, L. (1993) 'Numerisk analyse av reduserte luftutslipp i Norge', *SNF Working Paper*, 93: 108, Bergen: Stiftelsen for samfunns – og næringslivsforskning.

Håkonsen, L. and L. Mathiesen (1994) 'Integrasjon av skadevirkninger i generell likevekt', *SNF Working Paper*, 94: 76, Bergen: Stiftelsen for samfunns og næringslivsforskning.

Haley, S. (1994) 'Assessing Environmental and Agricultural Policy Linkages in the European Community', in J. Sullivan (ed.), *Environmental Policies: Implications for Agricultural Trade*, US Department of Agriculture, Foreign Agricultural Economic Report, 252, Washington, DC.

Hannesson, R. (1991) 'En samfunnsøkonomisk lønnsom fiskerinæring: Struktur, gevinst og forvaltning', *SNF-Arbeidsnotat*, 21/1991, Bergen: SNF.

Hanneson, R. (1993) *Bioeconomic Analysis of Fisheries*, Oxford: Blackwell Scientific.

Hansen, S. (1993a) *Er vi rede til nullvekst?: U-landssolidaritet med miljøprofil*, Oslo: Cappelen.

Hansen, S. (1993b) *Miljø- og fattigdomskrise i sør: Et utvidet utviklings-økonomiskperspektiv*, Oslo: Universitetsforlaget.

Hansen, S. (1994) 'The Market for Environmental Goods and Services', in *Financing Environmentally Sound Development*, Manila: Asian Development Bank: 285–322.

Hansen, S., P. F. Jespersen, I. Rasmussen and F. Theisen (1995) *en tilpasning av premissene – Fra indikatorer til krav* (An adaptation of the Premises – from Indicators to Demands) PBØ Report 5 (January), Oslo: Prosjekt Alternativ Framtid.

Hanski, I. *et al.* (1988) 'Populations and Communities in Changing Ecosystems in Finland', *Ecological Bulletin*, 39: 159–68.

Hartwick, J. (1977) 'Intergenerational Equity and the Investing of Rents from Exhaustible Resources, *American Economic Review*, 66: 972–4.

Heiberg, E. and K. G. Høyer (1993) 'Persontransport – konsekvenser for energi og miljø', Report 93: 1, Sogndal: Vestlandsforskning.

Hervik, A. and O. Hauge (1995) *Grunnrentebeskatning i et miljøperspektiv* (Economic Rent Taxation in an Environmental Perspective) PBØ Report, 13 (July), Oslo: Prosjekt Alternativ Framtid.

Hicks, J. (1946) *Value and Capital*, Oxford: Claredon.

Holm, S. and Holm, V. (1989) 'Arealavrenning fra jordbruket: En modellstudie med vekt på endret regional produksjonsfordeling og endret arealintensitet', *Report*, 12, SEFO, Ås.

Hulten, C. R. (1992) 'Accounting for the Wealth of Nations: Net Versus Gross Output Controversy and its Ramifications', *Scandinavian Journal of Economics*, 94: 9–24.

IPCC (1991) *Climate Changes: the IPCC Response Strategies*, Island Press.

Izrael, Y. *et al.* (1992) 'Final Draft Contribution to the IPCC 1992 Supplement', IPCC-VII/Doc. 4.

Jervell, A. M. (1994) 'Kvoter, kostnader og kapasitet i melkeproduksjonen', *Landruksøkonomisk Forum*, 4: 17–30.

Jespersen, P. F. (1995) *Mulige følger for persontransporten av en større CO_2-avgift* (Possible Consequences for the Transport Sector of Increased CO_2 Taxation), PSE Report, 6 (August), Oslo: Prosjekt Alternativ Frammtid.

Jorgenson, D. W. and P. J. Wilcoxen (1994) 'Reducing US Carbon Emissions: An Econometric General Equilibrium Assessment', in D. Gaskins and J. Weyant (eds), *The Costs of Controlling Greenhouse Gas Emissions*, Stanford: Stanford University Press.

Jorgenson, D. W. and K. Y. Yun (1990) *Tax Policy and The Cost of Capital*, Oxford: Oxford University Press.

Jorgenson, D. W., P. J. Wilcoxen and P. Pauly (1992) 'The Efficiency Value of Carbon Tax Revenues', US Environmental Protection Agency, mimeo.

Kielland-Lund, J. (1991) 'Diversitet i naturlige engsamfunn', Faginfo, *SFFL Report*, 23: 64–9.

Kolsrud, D. and E. Nesset, (1992) 'Sysselsettingseffekter av redusert arbeidsgiver-avgift', *Sosial økonomen* 6: 22–8.

Kverndokk, S. and K. E. Rosendahl (1995) 'CO_2-avgifter og petroleumsformue', *Økonomiske Analyser*, 1, Oslo: Statistics Norway.

Lagerlöf, J. (1985) 'Fauna och faunavård i jordbrukslandskapet', *Fauna och Flora*, 4–5: 149–59.

Lave, L. B. and E. P. Seskin (1977) *Air Pollution and Human Health*, Baltimore, MD: Johns Hopkins University Press.

Leuck, D. (1994) 'The EC Nitrate Directive and its Potential Effects on EC Livestock Production and Exports of Livestock Products', in J. Sullivan (ed.) *Environmental Policies: Implications for Agricultural Trade, Foreign Agricultural Economic Report*, 252, Washington, DC, United States Department of Agriculture.

Lone, Ø. (1992) 'Accounting for Sustainability: Greening the National Accounts?', paper presented to the CIDE Workshop on Environmental and Natural Resource Accounting, Nairobi (24–26 February).

Lothe, S. (1995) 'En analyse av endringer i relativ konkurransedyktighet i meierisektoren som følge av ulike scenarier for miljøreguleringer i EU', An Analysis of Change in Relative Competitiveness in the Dairy Sector as a Result of Various Scenarios for Environmental Regulation in the EU, PSE Report, 12 (August), Oslo: Prosjekt Alternativ Framtid.

Luhmann, H. J. (1995) 'Ecological Tax Reform – Presentation of the Debate in Germany and Policy Proposals for the International Society', paper presented at the Danish Board of Technology Conference on Ecological Tax Reform, Copenhagen (22 June).

Lurås, H. (1994) 'Grunnrente og formue av norske naturressurser', *Økonomiske analyser*, 94: 8 Oslo: Statistics Norway: 9–18.

Lyche, A. (1992) 'Kostnader i kornproduksjonen ved varierende bruks størrelse', Hovedoppgave (Graduate Thesis), IØS, NLH, Ås.

Mader, H. J. (1988) 'Effecs on Increased Spatial Heterogeneity on the Biocenosis in Rural Landscapes', *Ecological Bulletin*, 39: 169–79.

Magnussen, K. (1994) *Verdsetting av miljøgoder* – Spørsmål knyttet til praktisk bruk av miljøpriser (Valuing Environmental Benefits – Questions Connected with the Practical Use of Environmental Pricing), Project for a Sustainable Economy, PSE Report, 4, Oslo: Prosjekt Alternative Framtid.

Magnussen, K. and S. Navrud (1995) *Miljøverdsetting som grunnlag for miljø og ressursforvaltning*, (Environmental Evaluation as a Base for Environ-mental and Resource Management), PSE, Report 9, Oslo: Prosjekt Alternativ Framtid.

Mäler, K. G. (1989) 'Sustainable Development', *Working Paper*, Economic Development Institute, The World Bank, Washignton D. C.

Mäler, K. G. (1991) 'National Accounts and Environmental Resources', in *Sustainable Development, Science and Policy: Environmental and Resource Economics*, Oslo, NAVF: 1–15.

Manne, A. S. and L. Mathiesen (1994) 'The Impact of Unilateral OECD Carbon Taxes upon the Location of Aluminium Smelting', *International Journal of Global Energy Issues*, 6: 52–61.

Manne, A. S. and R. Richels (1994) 'The Cost of Stabilizing Global CO_2 Emissions: A Probabilistic Analysis Based on Expert Judgements', *The Energy Journal*, 15: 31–56.

Marchand, M. P., P. Piestieau and S. Wibaut (1984) 'Optimal Commodity Taxation and Tax Reform Under Unemployment', *Scandinavian Journal of Economics*, 91: 547–63.

Marglin, S. A. (1963) 'The Social Rate of Discount and the Optimal Rate of Investment', *Quarterly Journal of Economics*, 77: 95–111.

Mathiesen, L. (1986) 'Et modellformat for markedsanalyser', *MU-Report*, 86: 4, Bergen: Senter for Anvendt Forskning.

Mathiesen, L. (1988) 'Analyzing the European Market for Natural Gas', *Working Paper*, 1998: 37, Bergen: Senter for Anvendt Vorskning.

Mathiesen, L. (1990a) 'Markedsanalyser for viktige norske eksportprodukter', *Sosialøkonomen*, 4: 18–23.

Mathiesen, L. (1990b) 'Et modellformat for analyser av regionale markeder for gas og elektrisitet', *Working Paper*, 90: 38, Bergen: Senter for Anvendt Forskning.

Mathiesen L. (1991) 'Analyse av energibruk og CO_2-utslipp i norsk økonomi i år 2000', *SNF Report*, 91: 54, Bergen: Stiftelsen for samfunns- og nœringslivsforskning.

Mathiesen, L. (1992) 'Arbeidsgiveravgift og samfunnsøkonomisk lønnsomhet', in K. Boye and A. Kinserdal (eds), *Små og mellomstore bedrifter i Norge – en analyse av betydning, lønnsomhets- og kapitalforhold*, *SNF Report*, 92: 87, Bergen: Stiftelsen for samfunns- og næringslivsforskning.

Mathiesen, L. (1993) 'Analyse av endringer i norsk faktoravgifter', in C. Anderesen, L. Mathiesen and J. G. Sannarnes, *Indirekte skatter i Norge. En analyse med fokus på næringsvirksomhet*, *SNF Report*, 93: 52, Bergen: Stiftelsen for samfunns- og næringslivsforskning.

Mathiesen, L. (1995) *Sysselsettingsvirkninger av reduserte CO-utslipp* (Employment Effects of Reduced CO_2 Emissions), PSE Report 14 (August), Oslo: Prosjekt Alternativ Framtid.

Matthey, W., J. Zettel and M. Bieni (1990) 'Invertebres bioindicateurs de la qualité de sols agricoles', Nationalen Forschungsprogrammes, 'Bonen', Berne: Liebefeld.

Merriam, G. (1988) 'Landscape Dynamics in Farmland', *TREE*, 3: 16–20.

Mikkola, K. (1987) 'Förändringar av fjärilsfaunan i Finland i relation til biotopförändringar efter år 1950', *Ent. Meddr*, 55: 107–13.

Ministry of Agriculture (1993) 'Landbruk i utvikling', *Stortingsproposisjon nr. 8: (1992–3)* (Parliamentary White Paper), Oslo: Department of Agriculture.

Ministry of Agriculture (1994) 'Høringsutkast: Landbruksdepartementet's handlingsplan for bevaring og bœrekraftig bruk av biologiskmangfold', Oslo: Department of Agriculture.

Ministry of Environment (1992) 'Betänkande av kommissionen for övervaking av hotade djur och växter', *Kommittebetätankede*, 1991: 30, Helsinki: Ministry of the Environment.

Ministry of Environment (1984) 'Samlet plan for vassdrag', Hovedrapport, *Stortingsmelding nr. 63 (1984–5)* (Parliamentary White Paper), Oslo: Ministry of the Environment.

Ministry of Environment (1987) 'Om Samlet plan for vassdrag', *Stortingsmeding nr. 53 (1986–7)* (Parliamentary White Paper), Oslo: Ministry of the Environment.

Ministry of Environment (1994) 'Biologlsk mangfold', Miljøverndepartementets delplan: Høringsutkast, Oslo: Ministry of the Environment.

Mittenzwei, K., A. Huus and S. S. Prestegard (1994) 'Norsk landbruk og EU', *Report*, C-033-94, Oslo: NILF.

Moen, K. J. and I. F. Klynderud (1994) 'Jordbrukets kulturlandskap: Framtidig utvikling i jordbrukets kulturlandskap belyst ved grunnverdibereginger av arealkrevende produksjoner', *Melding, nr* 12, IØS, NLH, Ås.

Moum, K. (ed) (1992) 'Klima, økonomi og tiltak (KLØKT)', *Rapport* 92/3, Oslo: Statistics Norway.

Munro, D. A. and M. W. Holdgate (1991) *Caring for the Earth: A Strategy for Sustainable Living*, Montreal: IUNC, UNEP, WWF.

Nærings-og energidepartementet (Ministry of Industry and Energy) (1993) *Faktaheftet 1993: Norsk petroleumsvirksomhet*, Oslo: Ministry of Commerce and Energy.

Naturvårdverket (1990) 'Hotade arter. Långsiktig handlingsprogram', Solna: Naturvårdsverket.

Niskanen, W. A. (1971) *Bureacracy and Representative Government*: Chicago: Aldine.

NOK (1989) 14. *Bedrifts- og kapitalbeskatningen – En skisse til reform*, Finans og Tolldepartementet.

Nordhaus, W. A. (1991) 'The Cost of Slowing Climate Change: A Survey', *The Energy Journal* 12: 1, 37–65.

Nordhaus, W. and J. Tobin (1972) 'is Growth Obsolete?' in *Economic Growth*, National Bureau of Economic Research General Series, *No* 96E, New York: Colombia University Press.

Noss, R. F. (1990) 'Indicators for Monitoring Biodiversity: A Hierarchical Approach', Biological Conservation 4: 355–64.

NOU (1991) 'Norsk landbrukspolitikk. Utfordringer, mål og virkemidler', Hovedinnstilling, Oslo: Statistics Norway.

NOU (1992) Mot en mer kostnadseffektiv miljøpolitikk i 1990-årene. NOU 1992: 3.

NOU 1992:3 Mot en mer kostnadseffektiv miljøpolitikk i 1990-årene.

Opdam, P. (1989) 'Understanding the Ecology of Populations in Fragmented Landscapes', transcripts of the 19th IUGB Congress, Trondheim, Norway.

Parry, I. W. H. (1994) 'Pollution Taxation and Revenue Recycling', *Working Paper*, Economic Research Service, US Department of Agriculture.

Pearce, D. and G. Atkinson (1993) 'Capital Theory and the Measurement of Sustainable Development: An Indicator of Weak Sustainability', *Ecological Economics*, 8, 103–8.

Pearce, D., A. Markandya and E. B. Barbier (1989) *Blueprint for a Green Economy*, London: Earthscan.

Pedersen, U. (1994) 'Effektivitetskostnader ved beskatning', *SNF Report*, *No* 94: 26, Bergen: Stiftelsen for samfunns- og næringslivsforskning.

Pigou, A. C. (1938) *The Economics of Welfare*, London: Weidenfeld & Nicolson.

Pigou, A. C. (1947) *A Study of Public Finance*, London: Macmillan.

Pimentel, D. and A. Warneke (1989) 'Ecological Effects of Sewage Sludge and other Organic Waste on Arthropod Populations', *Agri. Zool. Rev.*, 3: 2–30.

Ploeg, F. van der and A. L. Bovenberg (1994) 'Economic Policy, Public Goods and the marginal Cost of Public Funds', *Economic Journal*, 104: 444–54.

Postel, S. (1993) 'Knappheten på vann', in L. W. Brown (ed.), *Jordens tilstand 1991*, Oslo: Aschehoug.

Postel, S. and J. C. Ryan (1991) 'Å reformere skogbruket', in L. W. Brown (ed.), *Jordens tilstand 1991*, Oslo: Aschehoug.

Prendergast, J. R., R. M. Quinn, J. H. Lawton, B. C. Eversham, and D. Q. Gibbons (1993) Rare species, the coincidence of diversity hotspots and conservation strategies, *Nature* 365: 335–7.

Prestegard, S. S. (1994) 'Norsk landbruk utanfor EU – kva rammer set EØS og GATT?', in R. Simonsen (ed.), *Norsk landbruk og EU – hva kan forskerne si?* NFR, Landbrukspolitisk forskningsprogram, Ås.

Proost S. and D. Van Regemorter (1998) 'Are there cost efficient CO_2 reduction possibilities in the transport sector? – Combining two modelling approaches', *International Journal of Environment and Pollution*, special issue.

Ramjerdi, F. and L. Rand (1992) 'The Norwegian Climate Policy and the Passenger Transport Sector', *TØI Report*, 152.

Rauscher, M. (1994) 'Trade Law and Environmental Issues in Central and East European Countries', *CEPR Discussion Paper*, 1045, London: CEPR.

Rawls J. (1971) *A Theory of Justice*, London: Oxford University Press.

Repetto, R., W. Magrath, M. Wells, C. Beer and F. Rossini (1989) *Wasting Assets, Natural Resources in the National Accounts*, Washington DC.: World Resources Institute (WRI).

Rickertsen, K. (1989) 'NAP – A Simulation Model for Norwegian Agriculture and Agricultural Policy', *Report*, 27/1989, Center for Applied Research, Oslo.

Rickertsen, K., Ø. Holand and S. Rystad (1995) *Skitten og deregulert? – Effekter på miljøe og ressursbruk av en internasjonalisering av jordbruket* 'Dirty and Deregulated' – The Impacts on the Environment and Resource Use of Internationalisation of Agricultures, Report 11 (August), Oslo: Prosjekt Alternativ Framtid.

Rijberman, S. R. J. (1990) 'Targets and Indicators of Climatic Change', in J. Jager (ed.), *Responding to Climate Change: Tools for Policy Development* (Summary Report), Stockholm: The Stockholm Environment Institute.

Roland, K. and T. Haugland (1995) 'Joint implementation', *EED Working Papers*, no. 7, Fridtjof Nansens Institute, Norway.

Romer, P. M. (1990) 'Endogenous Technical Change', *Journal of Political Economy*, 98: 71–102.

Roningen, V., J. Sullivan and P. M. Dixit (1991) 'Documentation of the Static World Policy (SWOPSIM) Modeling Framework', *ERS Staff Report*, 9151, Economic Research Service, United States Department of Agriculture.

Rønning, C. (1994) 'Analyse av regulering av luftforurensende utslipp med spesiell vekt på CO_2 – og Nox-utslipp', *SNF Working Paper*, 94: 75, Bergen: Stiftelsen for samfunns- og næringslivsforskning.

Rønningen K. (1994) 'Multifunctional Agriculture in Europe's Playground? Policies and Measures for the Cultural Landscape', Department of Geography, University of Trondheim.

Rose, C. and P. Hurt (1992) 'Can Nature Survive Global Warming?', *Discussion Paper*, London: World Wide Fund for Nature.

Ryan, J. C. (1992) 'Å bevare det biologiske mangfoldet', in L. W. Brown (ed.), *Jordens Tilstand 1992*, Oslo: Aschehoug.

Saunders, D. A., R. J. Hobbs and P. R. Ehrlich (eds) (1993) *Nature Conservation 3: Reconstruction of Fragmented Ecosytems*, Surrey: Beatty & Sons.

Sawyer, J. (1989) *Acid Rain and Air Pollution*, London: World Wide Fund for Nature.

Sefton, J. A. and M. R. Weale (1994) *The Net National Product and Exhaustible Resources: The Effect of Foreign Trade*, NIESR and Cambridge University.

Shackleton, R., M. Shelby, A. Cristofaro, R., Brinner, J. Yanchar, L. Goulder, D. Jorgenson, P. Wilcoxen, P. Pauly and R. Kaufman (1992) The efficiency value of carbon tax revenues, United States Environmental Protection Agency, Washington, DC., mimeo.

Shah, A. and B. Larsen (1992) 'Carbon Taxes, the Greenhouse Effect and Developing Countries', *World Bank Policy Research Working Paper*, Series 957, Washington, DC: World Bank.

Simonsen S., S. Rystad and C. K. Christoffersen (1992) 'Avgifter eller detaljregulering?: Studier av virkemidler mot nitrogenforurensning fra landbruket', *Report, No* 10, Institutt for Økonomi og samfunnsfag, NLH, Ås.

Solow, R. M. (1974) 'Intergenerational Equity and Exhaustible Resources', *Review of Economic Studies*: Symposium on the Economics of Exhaustible Resources, 41: 29–45.

Solow, R. M. (1993) 'Special Lecture: An Almost Practical Step towards Sustainability', *Resources Policy*, 19: 162–72.

Solow, R. M. and W. S. Vickrey (1971) 'Land Use in a Long Narrow City', *Journal of Economic Theory*, 3: 430–47.

Statens Forurensingstilsyn (1990) 'Tiltakskatalog for reduksjon av klimagasser i Norge', Bidrag til den interdepartementale klimautredning, Oslo: Norwegian Pollution Control Authority (SFT).

Statens Forurensingstilsyn (1991) 'NOx-analyse. Beskrivelse av utslipp i dag og frem mot år 2000', Oslo: Norwegian Pollution Control Authority (SFT).

Statens Forurensingstilsyn (1993) 'Forurensing i Norge: 1992', Oslo: Norwegian Pollution Control Authority (SFT).

Statistics Norway (1965) 'Norges økonomi etter krigen', *Samfunnsøkonomiske studier*, 12, Oslo.

Statistics Norway (1992) 'Naturressurser og miljø 1991', *Rapport* 92/1.

Statistics Norway (1993) *Statistik årbok 1993*, Oslo: Statistics Norway.

Statistics Norway (1994a) *Historisk Statistikk 1994*, Oslo: Statistics Norway.

Statistics Norway (1994b) *Nasjonalregnskapsstatistikk*, Oslo: Statistics Norway.

Steen, E. (1980) 'Dynamics and Production of Semi-natural Grassland Vegetation in Fennoscandia in Relation to Grazing Management', *Acta Phytogeogr. Suec*, 68: 153–6.

Stinner, B. R. and House, G. J. (1990) 'Arthropods and other Invertebrates in Conservation-Tillage Agriculture', *Annual Review of Entomology*, 35: 299–318.

Strand, J. (1992a) 'Environmental Policy under Labor Unions with Endogenous Environmental Effort', Memorandum from Sosialøkonomisk Institutt, 25, University of Oslo.

Strand, J. (1992b) 'Environmental Policy under Worker Moral Hazard', Memorandum from Sosialøkonomisk Institutt, 24, University of Oslo.

Strand, J. (1993a) 'Sysslsettingsvirkninger av vannkraft- og gasskraftutbygging i Norge: Generell analyse med eksempler fra Sauda, Heidrun og Tjeldbergodden;, Project for a Sustainable Economy, *Report*, 7, Oslo: Prosjekt Alternative Framtid.

Strand, J. (1993b) 'Business Fluctuations, Worker Moral Hazard and Optimal Environmental Policy', Memorandum from Sosialœkonomisk Institutt, 12, University of Oslo (published in H. Dixon and N. Rankin (eds), *The New Macroeconomics: Policy Effectiveness and Imperfect Competetion'*, Cambridge: Cambridge University Press, 1995).

Strand, J. (1994a) 'Environmental Policy, Worker Moral Hazard, and the Double Dividend Issue', Sosialøkonomisk Institutt, University of Oslo, mimeo.

Strand, J. (1994b) 'Employment and Wages with Sector-specific Shocks and Worker Moral Hazard', Sosialøkonomisk Institutt, 18, University of Oslo, mimeo.

Strand, J. (1995a) *Er miljøpolitikk sysselsettingsfremmende? En analyse av mulighetene for 'doble gevinster' gjennom skjerpet miljøbeskatning* (Can Green Policies Promote Employment? An Analysis of the Possibility of 'Double Dividends' from More Stringent Employment Taxation), PBØ Report, 6, Project for a Sustainable Economy, Oslo: Prosjekt Alternative Framtid.

Strand, J. (1995b) *Employment Effects of Hydropower and Gas Projects in Norway: An Analysis with Examples from Sanda, Heidruni and Tjellbergodden*, PSE Report, 7 (January).

Svensson, R. and M. Wigren (1986) 'A Survey of the History, Biology and Preservation of Some Retreating Synantropic Plants', *Symbolae botanicae Uppsaliensis*, 25.

Taylor, P. D., L. Fahrig, K. Henein and G. Merriam (1993) 'Connectivity is a Vital Element of Landscape Structure', *Oikos*, 68: 771–3.

Theisen, F. (1993) *Premissnotat for Prosekt Baerekraftig Økonomi*, PBØ, 1 (November).

Uhlen, G. and H. Lundekvam (1988) 'Avrenning av nitrogen og fosfor og jord fra jordbruk 1949–1979/88', *Report*, 7, SEFO, Ås.

UNDP (1992) *Human Development Report 1992*, New York: Oxford University Press.

UNEP (1992) 'Convention on Biological Diversity', United Nations Environment Programme, Nairobi.

UNEP (1993) *The World Environment 1972–1992*, pp. 406–7.

Vatn, A. (1984) 'Teknologi og politikk. Om framveksten av viktige styringstiltak i norsk jordbruk 1920–1980', Oslo: Landbruksforlaget.

Vatn, A. (1994) 'Estimerte avlingskurver for eng', Arbeidsnotat nr 27 fra ØkØk, Reassursforvaltning og forurensning i landbruket. Institutt for økonomi og samfunnsfag, NLH (Norway's Agricultural College) Ås.

Towards a Sustainable Economy Reports

Alfsen, K. Grønt BNP – trenger vi det?, December 1993.

Alfsen, K., Larsen, B. And Vennemo, H. Bœrekraftig Økonomi – Noen alternative modellscenarier for Norge mot år 2030, August 1995.

Asheim, G. Kompenserende investeringer i en ressursrik åpen økonomi, June 1995.

Farmer, M. Bærekraftighet i et økonomisk perspektiv, June 1995.

Hansen, S., Jespersen, P. F., Rasmussen I. og Theisen F. En tilpasning av premissene – fra indikatorer til krav, January 1995.

Hervik, A. And Hauge, O. Grunnrentebeskatning i et miljøperspektiv, July 1995.

Jespersen, P. F. Fremdriftsrapport for Prosjekt Bærekraftig Økonomi, March 1994.

Jespersen, P. F. Mulige følger for persontransporten av en større CO_2-avgift, August 1995.

Lothe, S. En analyse av endringer i relativ konkurransedyktighet i meierisektoren som følge av ulike scenarier for miljøreguleringer i EU, August 1995.

Magnussen, K. Verdsetting av miljøgoder – spørsmål knyttet til praktisk bruk av miljøpriser, 1994.

Magnussen, K. and Navrud, S. Miljøverdsetting som grunnlag for miljæ og ressursforvaltning, June 1995.

Prosjekt Bærekraftig Økonomi (PBØ) Rapport 1
Prosjekt Bærekraftig Økonomi (PBØ) Rapport 2
Prosjekt Bærekraftig Økonomi (PBØ) Rapport 3
Prosjekt Bærekraftig Økonomi (PBØ) Rapport 4
Prosjekt Bærekraftig Økonomi (PBØ) Rapport 5
Prosjekt Bærekraftig Økonomi (PBØ) Rapport 6
Prosjekt Bærekraftig Økonomi (PBØ) Rapport 7
Prosjekt Bærekraftig Økonomi (PBØ) Rapport 8
Prosjekt Bærekraftig Økonomi (PBØ) Rapport 9
Prosjekt Bærekraftig Økonomi (PBØ) Rapport 10
Prosjekt Bærekraftig Økonomi (PBØ) Rapport 11
Prosjekt Bærekraftig Økonomi (PBØ) Rapport 12
Prosjekt Bærekraftig Økonomi (PBØ) Rapport 13
Prosjekt Bærekraftig Økonomi (PBØ) Rapport 14
Prosjekt Bærekraftig Økonomi (PBØ) Rapport 15
Prosjekt Bærekraftig Økonomi (PBØ) Rapport 16
Rickertsen, K., Holand, Ø., and Rystad, S. Skitten og deregulert? – Effekter på miljø og ressursbruk av en internasjonalisering av jordbruket, August 1995.
Strand, J. Er miljøpolitikk sysselsettingsfremmende? En analyse av mulighetene for 'doble gevinster' gjennom skjerpet miljøbeskatning, December 1994.
Strand, J. Sysselsettings-virkninger av Vannkraft- og Gasskraftutbygging i Norge: Generell analyse med eksempler fra Sauda, Heidrun og Tjeldbergodden, January 1995.
Theisen, F., Premissnotat for Prosjekt Bærekraftig Økonomi, November 1993.

Author Index

Subject Index